Guadagno

The Art
of Science
Writing

THE
ART
OF
SCIENCE WRITING

Dale Worsley
&
Bernadette Mayer

Teachers & Writers Collaborative
New York

The Art of Science Writing

Teachers & Writers
Collaborative
5 Union Square West
New York, NY 10003

Library of Congress Cataloging-in-Publication Data
Worsley, Dale, 1948-
    The art of science writing.

    Bibliography: p.
    1. Technical writing—Handbooks, manuals, etc.
I. Mayer, Bernadette. II. Title.
T11.W68 1988   808'.0665   88-29569
ISBN 0-915924-20-X

Printed by Philmark Lithographics, New York, N.Y.

Third printing

Table of Contents

Acknowledgments

We wish to thank Phyllis McGrath, Manager of the Corporate Services Programs at the General Electric Foundation, and Nancy Larson Shapiro, Director of Teachers & Writers Collaborative, for their initiative in creating the General Electric Science Writing Project at the Manhattan Center for Science and Mathematics, where Cole Genn, principal; Harvey Kaye, dean of the Technology Department and director of the General Electric Scholar's program; Herb Laden, dean of the Science Department; Beatrice Ramirez-Epstein, dean of the English Department; teachers Glen Album, John Borchert, Joe Ciparick, Mitch Elinson, Jeffrey Levin, Stori MacPhee, and Joe Sherman; librarian Rita Moran; and Sandra J. Price provided invaluable assistance.

We would also like to thank our colleagues at the Institute for Writing and Thinking at Bard College: Paul Connolly, Director; Teresa Vilardi, Associate Director; Margaret Bledsoe and Alan Marwine, workshop leaders; and at Teachers & Writers Collaborative: Chris Edgar, Elizabeth Fox, Herb Kohl, Gary Lenhart, Pat Padgett, and our editor Ron Padgett. William Logan and Ron Padgett wrote part of the Writing Experiments section, and Russel Kenyon provided the appendix on math writing. Ann Chamberlain of the Exploratorium in San Francisco provided many helpful materials.

We would especially like to thank all our students for enabling us to learn more about the processes of science writing.

The following publishers and writers allowed us to quote from their material:

F. Woodbridge Constant, *Fundamental Principles of Physics*, copyright © 1967 Addison-Wesley Publishing Co. Pages 262-263 reprinted by permission.

Excerpt from "Abundance" by Peter Steinhart reprinted from *Audubon*, the magazine of the National Audubon Society. Used by permission of the author.

"Dreams of the Death of Beloved Persons" from *The Interpretation of Dreams* by Sigmund Freud, translated from the German and edited by James Strachey. Published in the United States by Basic Books, Inc. by arrangement with George Allen & Unwin Ltd. and The Hogarth Press, Ltd. Reprinted by permission of Basic Books, Inc., Publishers.

"Chemists on Mt. Olympus" by Joseph D. Ciparick reprinted by permission of the author.

"William James" by Martin Gardner from *The Sacred Beetle & Other Great Essays in Science* edited by Martin Gardner. Reprinted by permission of Prometheus Books.

Excerpt from *Chaos* by James Gleick. Copyright © 1987 by James Gleick. All rights reserved. Reprinted by permission of Viking Penguin Inc.

Material from Ruth von Blum's paper delivered at Bard College's 1987 conference on the Role of Writing in Learning Mathematics and Science reprinted by permission of the author.

Excerpt from Buckminster Fuller's "How Little I Know" © 1976 Estate of R. Buckminster Fuller used with permission of the Estate. For information about the work of Buckminster Fuller, contact the Buckminster Fuller Institute, 1743 South La Cienega Blvd, Los Angeles, California 90035.

We also express our appreciation to the following for their support of Teachers & Writers Collaborative, in addition to the General Electric Foundation: American Stock Exchange, Mr. Bingham's Trust for Charity, Columbia Committee for Community Service, Consolidated Edison, NYC Department of Cultural Affairs, Aaron Diamond Foundation, Long Island Community Foundation, Manufacturers Hanover Trust Company, Mobil Foundation, Morgan Guaranty Trust Company, Morgan Stanley Foundation, National Endowment for the Arts, New York Foundation for the Arts, New York State Council on the Arts, New York Telephone, New York Times Company Foundation, Henry Nias Foundation, Helena Rubinstein Foundation, The Scherman Foundation, and the Steele-Reese Foundation.

Preface

The Art of Science Writing is addressed to secondary school teachers who are interested in developing good student writing in the areas of science and math. Such teachers need not be science or math teachers—there is much here for English and social studies teachers as well. Our purpose is not only to bring creative ideas to the subject of science, but also to bring science ideas into the writing of essays, fiction, and poetry. Furthermore, many of the ideas in this book will prove useful to elementary and college level teachers, and to anyone interested in learning more about science writing.

The Art of Science Writing is an expression of our personal interest in science and mathematics as creative writers, and reflects the many years we have spent conducting writing workshops through Teachers & Writers Collaborative and other organizations. In 1986 and 1987 we worked in the classrooms at the Manhattan Center for Science and Mathematics, a New York City public school that admits students on a selective basis to put them on a science and technology track to college. We also attended workshops and conferences in math and science writing at the Institute for Writing and Thinking at Bard College.

Some of the ideas presented here are more fun than traditional competitive, statistics-oriented scientific pursuits, but our methods and suggestions can dramatically enhance any type of scientific inquiry. The ideas show respect for students, teachers, parents, and administrators (who may be working in systems that do not necessarily foster such respect). They are derived from the premise that writing in the classroom is essential to grasping the concepts of *real* science and math, that is, *as scientists and mathematicians know them.*

The Art of Science Writing offers many useful exercises for writing across the curriculum, whether the discipline be science, math, or anything else, for it is based on the fact that the boundaries between disciplines — between optics and psychology, logic and music, music and acoustics, astronomy and religion, mathematics and poetry, dance and ecology, etc.— are not definite.

It is not necessary to read this book from cover to cover. Instead, we advise readers to start with the section most suited to their needs. In some instances, the authors have taken different approaches to the challenge of writing in science and mathematics. Bernadette Mayer's ideas appear in the Writing Experiments section and Dale Worsley's in the Essay Development Workshop section. Occasionally there may appear to be contradictions. For example, Bernadette encourages reading freewriting aloud occasionally. Dale doesn't. The different authors are implying different uses for similar ideas.

Dale Worsley's Essay Development Workshop section leads the reader through the steps in the development of an essay. It uses procedures that are methodical and orderly, emphasizing logic, reason, and cogency. It takes into account the problems and frustrations that confront many teachers when they try to implement new programs into rigid curriculums.

Bernadette Mayer's Writing Experiments chapter presents many specific ideas for bringing out the inspired, intuitive, and associative side of the mind, and developing a sense of play and pleasure in the process of writing, so that writing can become both a tool and an art for aspiring scientists.

The Questions and Answers section is a forum for the discussion of issues that attend the uses of writing in science and mathematics. It answers the questions teachers often ask about teaching science writing.

The Samples section provides distinguished examples of writing in the sciences by students, teachers, scientists, journalists, philosophers, poets, and fiction writers. It is a good place to go for inspiration.

The annotated bibliography is necessarily eclectic, reflecting the books that we have found relevant, worthwhile, useful, or interesting for teaching science writing.

The appendix provides some special ideas for teaching math writing.

We advise that readers simply start with whatever section interests them most, go only as far as they like, and try out ideas only as long as they are engaging. We hope our book will stimulate the desire among students and teachers to write, to love to write, and to publish writing that is beautiful, speculative, and new.

Essay Development Workshop

GROUNDWORK

Science: a branch of knowledge or study dealing with a body of facts or truths, systematically arranged and showing the operation of general laws.

Essay: a short literary composition on a particular theme or subject, usually in prose and generally analytic, speculative, or interpretive.

Wisdom: 1. quality or state of being wise; knowledge of what is true or right coupled with just judgement as to action; sagacity, discernment, or insight. 2. scholarly knowledge or learning: *the wisdom of the schools.*

Verbal comprehension is helpful to the understanding of mathematical and scientific concepts, and verbal articulation is essential to the communication of the concepts. Writing is the medium by which wisdom is brought to science and mathematics.

Although some of us are in schools that have adopted good writing practices in the classroom, many of us — I venture to say *most* of us — are working in schools that have not. The lack of writing correlates with a great deal of frustration.

Some of the reasons writing is not used in math and science classes: 1) We don't have time to correct a lot of student essays. 2) We are too insecure about writing ourselves. It's easier to limit ourselves to dealing with the information of the subject. 3) If we go out on a limb and try to use writing in our classes, other teachers will be resentful and suspicious. 4) Writing should be taught by English teachers, not science and mathematics teachers.

Some of the reasons teaching is frustrating: 1) There is not enough time to treat interesting subjects in depth. 2) The textbook is crammed with laws, facts, equations, and formulas and the students' curiosity is dying while they try to memorize them. 3) If we stray from the curriculum to let the students ask other questions, they won't do well on the standardized tests. There is pressure to "teach for the test."

1

If you have experienced these hesitations and frustrations, perhaps you would find it illuminating to write about them before going on. Address this question: "What is it that makes meaningful teaching difficult?" Remember that spelling, grammar, and neatness are not important at this stage. In fact, worrying about them may inhibit your thinking. Please feel free to say anything you want in any way you like.

(Pause.)

One of the ubiquitous problems in education is communication between students, parents, teachers, administrators, politicians, and business leaders. Later in this section I'll talk about the importance of discussion. At this point, though, bringing up the idea of pausing to listen is pertinent. In an unpublished essay, Dr. Hassler Whitney, Professor Emeritus of Mathematics at the Princeton Institute for Advanced Study and a distinguished consultant on the teaching of arithmetic in the schools, emphasizes the importance of pausing: "In a group of a few people desiring to get into the heart of matters, there are some simple principles that help. Making short, carefully thought-out statements, and then *pausing*, promotes real listening and consideration. The others *and* you gain from this. Neither pushing nor pulling may help; but accepting the *process* underway—and looking for a time to continue it—is likely to give at least a few people the courage to go further." In pausing to write, we listen to ourselves and hear the promptings of our own minds. Listening to ourselves is where communication begins.

•

Before reaching to a tool chest, we should stand back and check the designs of what we propose to build. What principles will guide us? We are skilled professionals, who care about our work. We derive satisfaction from doing it well. We have ideals, morals, and a sense of ethical responsibility. While we support ourselves by teaching, the ultimate priority in the classroom is the good of the student. And while we know that what is good for them is good for ourselves, it is their future that concerns us most. If they are science and mathematics students, then they are, in fact, young scientists and mathematicians. To equip them for their futures in their fields we need to treat them with professional respect. Likewise, to flourish in our vocation we need to be accorded respect from the students, fellow teachers, and administrators. While some of us may be caught up in systems that threaten our ideals, we still cling to our principles, because without them, there will be little meaning in what we do.

2

Moreover, there is no chance of changing a poorly functioning system without first appreciating our own worth.

That is how I see it. You may see it somewhat differently. If so, please pause here to write down your ideas.

(Pause.)

Having reminded ourselves of who we are, and having confirmed our determination to build solid educational edifices, we find that the tool we need is already in our hands.

•

Some teachers question the idea that verbalization helps us understand scientific and mathematical concepts, but other contemporary thinkers confirm the value of such verbalization.

Marcia Birken, Assistant Professor of Mathematics at the Rochester Institute of Technology, Honors Calculus teacher, and recipient of the Eisenhart Award for Outstanding Teaching says, "The same skills are needed to construct a well-organized essay and to construct a logical proof."

William J. Mullin, from the Department of Physics and Astronomy, the University of Massachusetts, has stated, "The hard sciences and mathematics are often presented in ways described variously as quantitative, formal, rigorous, mathematical, etc. In spite of that correct description, the fundamental understanding professional physicists have of their field contains considerable elements that are qualitative, intuitive, and non-mathematical."

Theoretical astrophysicist David Layzer, from the Harvard-Smithsonian Center for Astrophysics, says that "while writing cannot replace the pictures you see with the eye of the mind, you can't learn to express yourself mathematically unless you can express yourself verbally."

Visualization is necessary for learning scientific and mathematical concepts. Although writing cannot replace it, visualization alone is not enough. Writing enables and enhances the intuition that science and math depend upon for understanding and progress. Good scientists and mathematicians write well.

Hungarian mathematician Janos Bolyai (1802-1860) said, "In the author there lives the perfectly purified conviction (such as he expects from every thoughtful reader) that by the elucidation of [his] subject one of the most important and brilliant contributions has been made to the real victory of knowledge, to the education of the intelligence, and consequently to the uplifting of the fortunes of men." Bolyai was referring to

3

the Euclidean and Hyperbolic Axioms of Parallelism, but the words could apply to many topics.

Naturalist John Burroughs says, in his essay "Science and Literature": "It is not my purpose to write a diatribe against the physical sciences. I would as soon think of abusing the dictionary. But as the dictionary can hardly be said to be an end in itself, so I would indicate that the final value of physical science is its capability to foster in us noble ideals, and to lead us to new and larger views of moral and spiritual truths. The extent to which it is able to do this measures its value to the spirit, measures its value to the educator. [. . .]Until science is mixed with emotion, and appeals to the heart and imagination, it is like dead inorganic matter; and when it becomes so mixed and so transformed it is literature."

Accepting these past and present authorities, we ask, "What should we do about it in the classroom?" It would be valuable to answer this question together. Please pause here to write your initial ideas on the subject. Let me pose the question this way: "What use does writing have in the science and mathematics classroom?"

(Pause.)

This is how the same question was answered by science and math teachers in a Bard College writing workshop:

To permit writing permits thinking. Writing can be used as a method of solving problems. It is a mirror of the mind to writers and a window to the mind for readers (sometimes, but not always, teachers), allowing both to see how well learning is taking place. As teachers *and* students, the more we write, receive responses to our writing, and react to the responses, the faster we grow in our knowledge, skill, and confidence. It is a way for *outsiders* (students) to become *insiders* in a new field. It can reduce anxiety by keeping thought flowing. It is a way for students to ask questions they might otherwise have been unable to ask. It captures elusive, but valuable, ideas. It not only can develop, but it can *create* knowledge on the part of the writer. It is a way to make and rectify mistakes, while curing the disease of thinking that mistakes are threatening and undignified. It is the currency by which people acquire ownership of ideas; ideas owned are ideas remembered, and ideas remembered are ideas learned. It is a way to bring students together into a trusting team. It is a method of slowing the pace of learning to ensure comprehension, which makes it more efficient, eliminating the need for desperate reviewing. It transforms boredom into curiosity.

Can writing *really* accomplish all that we claim? Yes, if applied in the right spirit.

•

While the writing procedures that we will examine in this cha
be used in any classroom regularly and effectively, they can be i
in classes free from curriculums geared to data-oriented te
writing suggestions have two general goals: 1) to produce short sketches
that assist learning in a variety of ways, and 2) to produce original, thoroughly researched and revised essays. The sketches can be considered
"process writing" or "microthemes" while the essays can be considered
"products" or "macrothemes." You might keep these distinctions in mind
as we go along. At the beginning, all of the techniques will be described
briefly, then each will be discussed in depth. I will refer to the writing in
the context of science, but it can usually be applied to mathematics as
well.

If you've been writing on scraps of paper so far, I suggest you now get a
small notebook to write in. It could prove useful as an example for your
students.

OVERVIEW

All of the steps that will be discussed in detail later are summarized
below, numbered 1-10. Because they are boiled down from my own practices as a writer, my experience in the schools, and what I learned from
colleagues, they reflect my particular situation. If you find them useful,
adapt them to your own habits and needs.

1. Exercises

Freewriting: A period of nonstop writing, perhaps three minutes, in
which all thoughts are jotted down as they occur. For the writer's eyes
only. Repeat as often as desired, usually at the beginning of each class.
Good for loosening up.

Directed Freewriting: A period of nonstop writing, perhaps ten
minutes, in which all thoughts on a particular subject or question are jotted down as they occur.

Discussion: All other activities stop for a limited period of time in
which the class addresses a problem, analyzes a topic, or debates a point.
Monitored by the teacher or an appointed student.

Summary Writing: A period of writing, perhaps five minutes, at the
end of a class or activity. May include ideas and points worth remembering, questions in need of answers, and plans for future action.

Visualization: A period of writing, perhaps five minutes, in which the
students, after considering an example from literature, use metaphor and
simile to describe an object or process.

Objective Observation: A period of writing, perhaps five minutes, in which the students describe an object or process as accurately as possible, eliminating preconceptions.

2. Inspiration
The presentation of an interesting writing sample that has inspired the teacher or a student. May occur at the beginning or at any appropriate point throughout the development of the essay. The example can be used to convey exciting scientific ideas or illustrate points about writing, such as the power of descriptive words or the necessity for factual substantiation.

3. Topic Selection
Students choose their essay topics. They may find good topics in a variety of ways: by reading textbooks, library books, and magazines, by discussing ideas with each other, by brainstorming, etc.

4. Topic Evaluation
Through directed freewriting, discussion, and summary writing, students get a better idea of whether or not their topics will be interesting to pursue. The evaluation is in itself the first step in the writing of the essay.

5. Research
Students find information related to their topic and spend time in class writing down their reactions to the information.

6. Rough Draft
Students assemble their thoughts and notes as coherently as possible.

7. Peer Criticism
Students trade their rough drafts or present them to the class as a whole for respectful analysis.

8. Revision
More drafts are written until the essay is fully developed to the author's satisfaction.

9. Publication
The essays are published and/or read to an audience, which may be the class.

10. Evaluation
Students write evaluations of the process of developing their essays so the teacher can know what they think of it.

If you see science writing as a creative act, as I do, then perhaps you also share my belief that creativity is not so much the generation of something that isn't there, as it is the removal of obstacles to the realization of something that has the potential to be there. The following exercises remove such obstacles by helping young writers determine what they want to do, know what they are doing while they are doing it, discover what they have done when they are finished, and figure out how to change it. The first four exercises are useful primarily as procedures, the last two I created to treat specific needs. (For a variety of other exercises, see the Writing Experiments section.)

Freewriting

For a limited amount of time, perhaps three to five minutes at the beginning of a class, students and perhaps teachers write down nonstop whatever comes into their minds. The thoughts needn't be complete; scratching out or making corrections is unnecessary; spelling and grammar are irrelevant.

I have found that freewriting works best when all the students are seated and have clear desks but for a blank sheet of paper and a pen or pencil. They should write no matter what their thoughts may be, keep writing without stopping, and put the writing away immediately when the time is up. The exercise should be started and stopped strictly on time. Counting down ten seconds to start it and informing the writers when thirty seconds are left before stopping it are good strategies. If students say they don't know what to write, they should be told to write precisely that: "I don't know what to write," because that is what is in their minds. A dramatic way to make the point clear is to say that the only acceptable reason for not writing is death, since there is always *something* in the mind. Even in sleep there are dreams. No talking should be permitted during the freewriting. Students must put the writing away immediately when it is done, and teachers should move on to the next step in the class quickly, before the freewriting assumes too much importance.

When students start a class with freewriting, they feel good about writing and appreciate you for giving them the chance to do it. Since they are permitted to have their own thoughts in the room, the physical space belongs to them and they are more willing to act responsibly. In essence, one begins with a pause. I use freewriting regularly, but skip it sometimes when it seems the class is already focused, motivated, and quiet.

Take five minutes here to try it. Time yourself exactly.

(Pause.)

(Please see "Freewriting" in the Writing Experiments section for another approach and an example.)

Directed Freewriting

For a limited amount of time, perhaps ten minutes, writers jot down their thoughts on a particular subject or idea. The thoughts are written as they occur and without correction. It is a way of focusing the mind and raising pertinent questions. It is useful particularly at the beginning of lessons and projects, or new phases of ongoing projects. It makes a writer's thoughts concrete enough to analyze, without the pressure of writing a full essay. Used in combination with discussion and summary writing (discussed below), it is an ideal way to see objectively what has been learned and what new thinking is taking shape. Once a teacher becomes discerning about when and where to use it, it is likely to become an integral part of her teaching.

Please take ten minutes to write your thoughts on a particular subject. I suggest the subject of evolution, since I use it as an example later.

(Pause.)

Discussion

Essays in the professional world are not simply reportage, they are created at the vortex of various currents of thought and activity. Whether through scientific journals or in the halls of a conference, ideas, opinions, and information of all kinds flow from mind to mind: they are discussed. As J. Robert Oppenheimer put it (in a lecture quoted more extensively in the Samples section), "The work of science is cooperative; a scientist takes his colleagues as judges, competitors, and collaborators." Discussion is the medium of cooperation among students as young scientists, as well. They think a lot and need to exchange ideas as much as professionals do.

One of my students put it this way: "I think instead of having a writing workshop, they should have a talking workshop where the students can discuss their feelings about the environment they live in and the problems teens of their own age go through. I think that before you get down and write your thoughts on a piece of paper you have to know what you want to think about."

By itself a discussion is not necessarily an exercise, but in combination with directed freewriting and summary writing it becomes one. Discussions can be oriented in a variety of ways. They may be about the ideas of a single student. They may be about a scientific, ethical, or moral issue. They may be about a classroom problem. Whatever purpose a discussion serves, certain standards apply: people should listen and speak with respect, individuals should not be allowed to dominate, shy people should be drawn out, and the discussion should not continue beyond its usefulness in exploring a subject.

Studies have shown that girls are often given less time to speak in classrooms than boys. The discussion leader should be aware of this and channel the discussion accordingly. I have noticed that students have difficulty recapitulating the points made during a discussion. Instituting formal recapitulations could help solve this problem.

It is often only after the regular use of discussion that some students begin to be stimulated. One physics student, who hadn't been interested in writing and who listened through several discussions and seemed bored, raised his hand and asked, "What is the difference between an idealist and a realist?" If he hadn't been able to speak and ask the question, he might've stayed bored, but when his question became a point of discussion, he began to see it was interesting, and wrote an essay about it.

Discussions also give feedback to the teachers, who naturally have misconceptions about what students think. In an electronics class of the technology department at New York's Manhattan Center, we had already covered material about ideals in science and I felt satisfied the students had a conception of how they applied. In a later meeting, we were talking about the judgments that engineers have to make in the course of their work. Somehow the discussion came around to the dilemma of the nuclear physicists who were asked to build the first atomic weapons. I asked, "Why would a moral man like Robert Oppenheimer help build the bomb?" Students answered, "For money." "For prestige." "For research." They expressed no idea of the moral conflict involved, and I realized they had not bridged a gap between their own idealistic instincts and their concept of the way scientists must work in the world. I decided it was important to talk more about the issue, and doing so paid off with essays that contained moral and ethical dimensions as well as scientific and technical ones. I would never have known to do this without a discussion.

Astrophysicist David Layzer noted that people remember eighty percent of what they say and twenty percent of what they hear. Students must speak in the classroom.

Summary Writing

For a limited amount of time, perhaps five minutes, at the end of an activity or class, students jot down significant observations, memorable points, questions in need of answers, and plans for future action. If there is no rush, the students may be able to frame their thoughts thoroughly. If there is a rush, as there is in the forty-five minute classes at many schools, summary writing may still prove useful as a net to catch ideas that might otherwise slip away, and a bridge to the next session. Some of the students' greatest discoveries will occur during summary writing and appear in their essays.

Here is an example of how directed freewriting (on the subject of evolution), followed by discussion, followed by summary writing can work:

> *Directed freewriting:* Evolution to me really doesn't exist. My feelings towards it are none. I feel that some superior being put us on this earth, that superior being meaning "God," to many a supervisory state of mind, to others a feeling of respect. I don't believe man came from a non-existing cell. It is said that the earth before this time was a dead structure, so how can something like life come along from something that has no life itself?

> *Summary writing after discussion:* Evolution didn't evolve from some non-living species. These are my opinions, but scientific studies and observations have put a bit of doubt in my mind. Maybe humans derived from apes and monkeys, but then again, they are very different in culture, appearance, and intelligence. They are somewhat intelligent but still have not reached the brain's ability to suck in more and more information. Humans have not yet done that either, but they are more civilized and in control of their daily behavior.
> — *Wanda Lucca, 10th grade*

In my summary writing, I have made observations such as, "I've got to slow my expectations and put them in perspective. Students' minds need to be focused gradually." I've also made notes for future reference: "Next: They are supposed to be bringing in rough drafts next week. If they don't, have them write them in class. Discussion format is good. Try breaking it into smaller groups and floating?"

Visualization

Clear visualization is a key to learning and understanding. Writing can not only indicate exactly how clearly one understands a concept, but help clarify the image of the concept. Good tools to express visualization are metaphor and simile.

Students in a couple of my classes had problems visualizing certain processes, not for lack of imagination, but because the curriculum was too rapid for them to stop and grasp the facts. When one of the exercises didn't succeed the first time, I repeated it, to make sure the students had the experience of visualizing clearly. I used two examples, Shakespeare's sonnet 130 and an excerpt from a science article in the *New York Times*.

The sonnet used metaphors in an ironic way to serve objectivity. Before reading the poem aloud, I had the students compare the parts of a lover to other things, as Shakespeare did. (He was using the classical poetic device of listing or cataloguing. When one lists one's lover's qualities, the poem is known as a *blazon*. See the *Teachers & Writers Handbook of Poetic Forms* for more information.) This is what my students came up with:

eyes—black onyx
lips—rose
breasts—cotton
hair—silk
cheeks—peaches
breath—peppermint

Then I read the sonnet:

My mistress' eyes are nothing like the sun;
Coral is far more red than her lips red;
If snow be white, why then her breasts are dun;
If hairs be wires, black wires grow on her head.
I have seen roses damask'd, red and white,
But no such roses see I in her cheeks;
And in some perfumes is there more delight
Than in the breath that from my mistress reeks.

I love to hear her speak, yet well I know
That music hath a far more pleasing sound;
I grant I never saw a goddess go;
My mistress, when she walks, treads on the ground:
And yet, by heaven, I think my love as rare
As any she belied with false compare.

The sonnet is a plea for truth, an appreciation of love, and a quite comical stand for objectivity. Students like it.

Here is the second example I used, a passage of descriptive writing by Walter Sullivan on recent discoveries of the nature of the center of our galaxy, the Milky Way:

A voyager to the heart of the galaxy would find the sky almost blindingly bright with stars.

According to recent observations by the Very Large Array of radio telescopes in New Mexico of the area within 200 light years of the Milky Way's center, the traveler would enter a region of gigantic, parallel gaseous filaments, arcing around the core, each about 100 light years long. Additional clouds reaching from there toward the core are cut by strange, narrow "threads."

11

Scientists suspect that this material is in violent motion, but whether it is falling into the core, circling it or being ejected has not yet been determined.

Infrared radiation recorded from closer to the core has shown it is surrounded by a doughnut-shaped cloud of dust and gas tilted slightly from the galaxy's plane. The cloud, about 12 light years in diameter, is heated by energy equal to that from 10 million Suns, but the source is uncertain.

The inner region of this cloud is moving so rapidly that some astronomers believe it is in the grip of gravity from a black hole equal in mass to that of several million Suns. Material squeezed, and consequently super-heated, as it falls into such a hole could heat the doughnut.

Another proposed energy source is radiation from millions of hot young stars formed in a "starburst" 10 million years ago. The explosions of such stars as their brief lifetimes ended could explain why the gas as far as 10,000 light years from the center is moving outward at high velocity.

Observations by radiotelescopes have shown that at or very close to the point around which the entire system rotates is the galaxy's most powerful source of radio waves. More recent observations have shown that the primary emissions are coming from a region smaller than that enclosed by the orbit of Saturn. The radiation is typical of that generated by wildly gyrating electrons, rather than that of a heat-producing star and some believe the black hole lies within this area.

The article provides many examples of the way language can express clear visualization. Dramatic adverbs and adjectives such as "blindingly" and "violent" appeal to our senses. Comparisons such as "doughnut-shaped" allow us to form accurate pictures. When I read it to one class, you could hear a pin drop. Here are some of the results when tenth-grade students described the processes of electrostatics and transistors:

> Electrostatically charged electrons are like babies ready to be born. Imagine a baby that wants to come out and is kicking. Like the electrons, it is vulnerable and susceptible to anything. —*Shawnel Boone*

> Electrostatics is the exchange of charges between two objects. Imagine a dance. Five guys are dancing with their girlfriends. If five girls come to the girls' side, they will not attract. Electrostatic activity is produced when opposite charges touch. Say guys are negatively charged and girls are positive. Two girls will not attract so no charge is changed. Two guys either. But a guy and a girl do attract. When these two charges touch, it creates a small flow of electricity. —*Rhea Pulido*

> A transistor is a speeding train at rush hour. It transports multitudes of people from one station to another, allowing them to go to work. The people are electrons. —*Anthony Babeca*

A transistor is as small as a pea and as important as a bridge. It works as a bridge by letting in the cars and making them leave faster than they came in. It has one way in and can let the cars leave in two directions, either to the highway or to the street.   —*John Paul Rivera*

Think of a cylindrical shape and cut it in half from top to bottom. Place the round half so that it looks at you. There are three wires as thin as the thinnest stems in a leaf at the bottom.   —*Edgar Tantigua*

Please pause here, if you like, to note your thoughts about this exercise.

(Pause.)

Objective Observation

It is difficult getting students to make objective observations when the students know how an experiment is supposed to come out. They may also have preconceptions that stand in the way of objectivity in their essays. To perceive something accurately is a matter of separating observation from preconception. Here is an example of a student's writing that could benefit from more objectivity:

If I were to write a letter to alien life forms, this is what I would say:
Greetings, aliens. I am an Earthling and I come from Earth. I came here to explain my existence and how I came to be.
It all started millions of years ago. A devastating explosion called the "Big Bang" brought all the planets together. By chance, up until now, our planet was the only one that housed living organisms that advanced throughout the ages. My race was once said to be an inferior race of sub-humans called apes. This, of course, was all a theory. Maybe we can collaborate all our knowledge and create a superior race that can rule throughout the universe.
Together, Alien and Human can reach the end of the cosmos.

Among the several preconceptions in the piece is the notion that apes are "sub-human." Objectively, apes' evolution from a common ancestor is parallel to human evolution, and they are different, perhaps less intelligent, but not, objectively speaking, "sub" human.

To address the need for objectivity, I passed out a sheet of paper that defined objectivity (free from personal feelings or prejudice; based on facts; unbiased) and observation (the act or instance of viewing or noting a fact for some scientific or other special purpose). The sheet also contained an assignment, to write objective observations, perhaps about themselves or their immediate environment. They were permitted to use any kind of language they felt appropriate, but I reminded them that metaphor and simile were good tools. I also reminded them that it was impossible to make mistakes at the beginning.

13

Here are some of the results.

There is this big black unknown machine in the front of our English room. I wonder what it is. There are a bunch of knobs and plugs piled monotonously on it. There is a big meter. They look like two power meters on a radio. This machine looks very mysterious and it looks like a machine that Frankenstein would have used to make the monster. It is a very mysterious machine.

—*James Amber, tenth grade*

The expressions of the hands at the party were energetic. The hands of one person were like a dog wagging his tail wildly. The people's hands were waving in the air like a bird when it is flying. Then hands were moving with the help of the music. —*Bernell Hollis, tenth grade*

In the computer classroom at least 20 or 30 machines move, with the glare of one machine in the face of another. The keyboards tick. Their faces are blank, as if their hearts were part of the machines. —*Victor Wright, tenth grade*

Please pause here to write your own observations about the use of exercises to address specific problems in the classroom.

(Pause.)

INSPIRATION

Because of the curriculums in many schools, students have a tendency to see the great scientists of the past as dead dictators of laws, and resent them for their monumental immobility, rather than appreciate them for the intellectual life they have breathed into history. Students need the liberation of myths, not the oppression of authority. They need to get a sense of the excitement of discovery felt by their forebears. They need to be inspired.

A key to the inspiration of students is the inspiration of the teacher. We pass along not only the idea, but the attraction to ideas. In the long run, that is likely to be the more important gift to the students. Many of us are afraid to show our enthusiasm for ideas. We are afraid to reveal ourselves. We fear making mistakes, or not knowing answers. We are expected by everyone, including ourselves, to be walking answer machines. Such expectations destroy the possibility of inspiration.

It is our enthusiasm, our willingness to reveal ourselves, that brings Aristotle, and Freud, and Euclid, and Galileo out of the books and museums to breathe. To hear the voices of these figures, the students have to be open enough to listen. They have to be curious. To be curious they have to be personally involved with the idea at hand. One good way to

14

get them involved is by writing. The student does not need to have studied the particular subject to begin.

You are acquainted with the key thinkers in your subject and can most easily find inspiring passages from their writing. Perhaps there is also an appropriate passage in the Samples section. Let's say that you are teaching a tenth-grade biology course that includes information about the principles of heredity, the structure of chromosomes, the processes of meiosis and mitosis, etc. Any of several important scientists could be drawn upon for inspiration at the beginning of this course. You might be interested in Mendel. You might want to quote a passage of his writing and talk about how the field of biology was in its infancy during his lifetime, how institutions such as the church supported intellectual exploration whereas now a college or corporation would support it. If not by Mendel, you might be intrigued by the intensely competitive atmosphere surrounding Watson and Crick's research into the structure of DNA, and want to quote from their original essay on the subject, or from *The Double Helix*. (Another controversial aspect of the Watson-Crick discovery, brought out in *Mothers of Invention* by Ethlie Ann Vare and Greg Ptacek, is their apparent denigration of a woman contributor to the experiments, Rosalind Franklin.) My particular interests made me want to go to Darwin. It seemed to me that the political and religious turmoil about the issue of evolution could be used as a source of energy to fuel the students' curiosity.

To get students involved, we need to define the salient ideas briefly for them. In the case of my biology students, I made sure they understood rudimentary definitions of evolution, and were acquainted with the notions of natural selection, the struggle for existence, and, though Darwin himself never used this terminology, survival of the fittest. After I defined these terms, we spent ten minutes of directed freewriting on the subject. I made sure to tell the students their writing would be read aloud later. (It is important to advise students beforehand when their writing will be seen by others.) I wrote along with them. If you haven't already done so, take the time now to write ten minutes of directed freewriting on our example topic of evolution.

(Pause.)

When the time was up, I collected the students' papers and read them aloud, along with my own paper. I called for reactions to each paper, but didn't let discussions go too long. There was a great deal of enthusiasm for the ideas expressed in the directed freewriting. Definite attitudes emerged both for and against the theory. The students discussed their religious beliefs, their moral attitudes, and their feelings. Some didn't feel dignified about being "descended from apes." They expressed doubts that this could be true. (For an example of fluctuation between doubt and belief, see tenth-grader Wanda Lucca's directed freewriting and summary writing pieces on page 10.)

Having cultivated the students' curiosity through directed freewriting and discussion, I read these paragraphs from two chapters of the *Origin of Species*:

> In looking at Nature, it is most necessary to keep the foregoing considerations always in mind — never to forget that every single organic being around us may be said to be striving to the utmost to increase in numbers; that each lives by a struggle at some period of its life; that the heavy destruction inevitably falls either on the young or the old, during each generation or at recurrent intervals. Lighten any check, mitigate the destruction ever so little, and the number of the species will almost instantaneously increase to any amount. The face of Nature may be compared to a yielding surface, with ten thousand sharp wedges packed close together and driven inwards by incessant blows, sometimes one wedge being struck, and then another with greater force.
>
> It may be said that natural selection is daily and hourly scrutinizing, throughout the world, every variation, even the slightest; rejecting that which is bad, preserving and adding up all that is good; silently and insensibly working, whenever opportunity offers, at the improvement of each organic being in relation to its organic and inorganic conditions of life. We see nothing of these slow changes in progress, until the hand of time has marked the long lapses of the ages, and then so imperfect is our view into long past geologic ages, that we only see that the forms of life are now different from what they formerly were.
>
> Although natural selection can act only through and for the good of each being, yet characters and structures, which we are apt to consider as of very trifling importance, may thus be acted on. When we see leaf-eating insects green, and bark-feeders mottled-grey; the alpine ptarmigan white in winter, the red-grouse the color of heather, and the black-grouse that of peaty earth, we must believe that these tints are of service to these birds and insects in preserving them from danger.

I pointed out what I thought was interesting about the passages, explained why the concepts were revolutionary, and expressed my enthusiasm for the poetry in the descriptions. The students' curiosity was already piqued because they had invested their own thought in the subject and contributed their ideas to the class. Though the students knew their writing was only intended to reflect musing, not perfect knowledge, and that the initial probings of the mind should not be compared to the developed observations of Darwin, there was a sense that they shared something with him. They knew their opinions mattered as much as his, right or wrong, and together, as a class, through discussion, they could pursue the truth. They knew there were no tricks, that while the consequences of their thoughts were potentially serious, they were free to turn an idea around playfully until they could make up their minds about the truth.

There wasn't enough time left that day to continue a discussion of Evolution, so we moved directly to five minutes of summary writing. (Again, please refer to Wanda Lucca's summary writing earlier in this section for an example.)

Please spend five minutes writing in your notebook about any questions you may have or whatever other thoughts seem pertinent to you at this stage. (Afterwards, you may find the "Mutual Aid" experiment interesting in light of the example about evolution.)

(Pause.)

The process of including students in the sensation of discovering original ideas can be carried from this stage into either of two directions. If we need to return to our normal lesson plans in order to cover the content of our curriculum, we can simply use the inspiration material whenever it seems appropriate along the way, to show how the information in the textbook fits into a larger historical and scientific context. The students' instincts to learn should be sufficiently unlocked to make them more receptive and comprehending than they otherwise would have been.

If we are going to write "macrotheme" essays, however, students might have to do further writing to get inspired enough to tackle a longer essay.

We have seen how the founders of science can speak directly to the students' sense of themselves. To develop a really thorough, meaningful essay, we should return to this idea of the sense of the self. People write best about what captures their interest, and usually they find themselves to be a topic of interest. The more they have been taught to respect, trust, and rely upon themselves, the more confident they are about exploring the world and including it in a consideration of themselves. Picture a baby looking over its shoulder to affirm the steadfastness of its parents before it crawls down the porch steps. A self is as large as the world one is willing to incorporate into it.

High school students are in a position to reach out and grasp a considerable amount of interesting and pertinent information. It is at their fingertips in newspapers, in the library, in the classroom. If they have a strong enough sense of themselves, they will seek out this information and write about it avidly. If they are confident enough, they will render it into concepts, develop it along different lines. If they are given enough room, they will find it natural to move from hypothesis to proposition to theory to proof and back, inductively and deductively, like scientists. They will write effectively, mingling fact, opinion, speculation, example, and observation, provided they are allowed to follow the line of development that interests them most — starting with themselves. From the footing of themselves, they can build a strong bridge (an essay) to another shore (a topic).

Building the footing for this bridge is simple: have the students write about themselves for ten minutes. Let them use whatever terms they like, but if they need suggestions, tell them to describe themselves objectively. If they need further clarification, tell them a brief chronological biography will do, or a description of their fears and desires, or the information they would tell a visitor from another planet. Any approach is appropriate. The effect will be much more inspiring if you, the teacher, also write about yourself. Perhaps you would like to take the time now to try it out. (You will be much better prepared to answer the questions of students if you encounter problems first and solve them by experience.)

(Pause.)

When the ten minutes are up, read the papers, or have the students read them. Here are some examples from the Manhattan Center for Science and Math. (For another approach to writing about oneself, see the experiment called "The Question of One's Relationship to the Universe.")

> I consist of an id, ego, and superego. I am hiding beneath my fears, hopes, and hatreds. Sometimes I peek out and surprise myself. I am shy and an extrovert, if that is possible. I guess you may call me an "ambivert." I don't like to feel intimidated. I fear the dark, for my imagination uses it for a playground to run amok. Shame fills me when I think how I have no control of my mind, but it does make life interesting.   —*Evelyn Peña, twelfth grade*

> This indeed is quite difficult, writing about myself, but due to strength both mentally and physically within me I shall continue and try my best.
> I'd like to start off by writing about my personality. I feel that it is a very strong point to my social window, which people see me through. I am very quiet until you get to know me. Depending on what type of person you are I will in good time find out more about you. I am very curious. I have a very

high reading and math I.Q. although these do not reflect my school work and scores greatly due to my laziness at times. I am an adventurous person. I like all type of foods, music, people, etc.

I always react to everything that happens in my life. I seldom show it, and keep it to myself at times. And I always try to keep it in mind so that I may refer back to it in the future.

Because of certain things I worry about I often react more seriously and more thoughtfully to them. Like, for example, when Reagan sent over the F-15's to Libya I began to worry and started bringing up conversations about World War III. —*François Belizaire, twelfth grade*

I hope that you—and your students, if you have been working with them concurrently with your study of this chapter—are now inspired and ready to proceed to leading them to their essays. If you are not, please stop here to write your doubts and fears. If you disagree with my approach, modify it to your satisfaction. In any case, pause to spend five minutes to summarize your thoughts about the process of inspiration.

(Pause.)

TOPIC SELECTION

As teachers our objective is to provide an environment that encourages students to choose to write about scientific subjects on their own. If they elect to write about something else, however, they should not be discouraged. Should they prefer to continue writing about themselves, it is because they still need to. Our purposes would only be defeated if we thwarted this need and had them write about a subject that didn't interest them. They would merely adapt to our demands for a style of behavior, not cultivate their curiosity and foster genuine learning and growth. And who is to say the self is not a fit subject for scientific inquiry?

While they will select their own topics, there are a great many ways we can guide them in their decisions. We can help them find an interesting line of development from their writing about themselves. We can encourage them to go to classmates for suggestions. We can conduct brainstorming sessions to generate lists of ideas. We can lead them to magazines, newspapers, and books for inspiration. Perhaps the textbook in our courses can provide a topic idea. We may believe that a specific subject will stimulate an individual student and want to direct him to that subject. Maybe the student will find it absorbing. If not, he should be encouraged to look elsewhere.

Every class will have a different opinion as to what are good topics. One group of my students was preoccupied with AIDS protection, mandatory drug testing, surrogate motherhood, terrorism, violence in sports, and teen sex and pregnancy. The social and ethical dimensions of these issues made them that much more viable as topics. Most students are idealistic and have an acute sense of justice. There are good reasons to tap this source of motivation—or, to state it more accurately—to *permit* the students to tap it themselves. Doing so allows science to stay in the natural relation to values and politics that it has in the real world. No scientific endeavor takes place in ivory tower isolation from social and political events.

Aside from public issues, students have specific personal interests that can become topics. One of my students had a close relative who had recently died of a drug overdose, and wanted to find a scientific slant to understand better what happened. Another's father benefitted from a heart pacemaker, and she had already begun asking questions about how it worked. These interests were absorbing enough to take the students a good distance into scientific research and inquiry. A student who chose artificial limbs as her topic but didn't have any personal experience with them quickly lost interest.

Normally we can expect success finding topics among current events and personal interests, but there are a couple of other circumstances that would make us look for other strategies.

If we are working in schools or with particular classes in which the students are deeply suspicious, resentful or repressed, and there is no sense of teamwork, no possibility of having successful discussions, no chance that the students will feel free enough to write about themselves, then we might want to use subterfuge to get them interested in writing. When I was dealing with such a class, where students were reluctant to speak out except to make disrespectful remarks, it occurred to me that they were feeling insecure. Doubtless there were many reasons why, but certainly one was that I was asking them to think openly, in public, when they were more familiar with assignments and tests geared to absorbing and regurgitating information on paper. Paradoxically, though these assignments and tests were a source of their lack of self-confidence, using the *forms* of this method of teaching made them feel more secure. Tests were the devil they already knew, so to speak, so I handed out a paper that looked like a test, but was, in fact, an inducement to write what they considered most interesting. Because no psychologically normal student is averse to learning, the cause for the lack of curiosity on the part of these students must have come from their "nurture," as opposed to their "nature," so I included social and political questions on my mock test. (Look

20

under "A Mock Test" in the Writing Experiments section for a list of the questions I handed out.) The responses were gratifying, and the students gradually gained the confidence to pick topics that genuinely interested them, some from the "test" questions, but many from other sources, which indicated to me that they were making real progress.

Another circumstance that requires a different approach is the opposite of the above: when students are already exceptionally motivated. Nothing but our ignorance of their desire will prevent some students from writing effective science essays. When I have worked with such students, listening to them carefully has helped solve their problems. One student did not want to write in class when the other students worked. I simply told him that was all right. He daydreamed while the other students wrote. They were enjoying their writing too much to feel as though he were getting away with something. He later produced a dazzling essay, on his own time. Another student was in a peculiar situation: the teacher working with me had assigned group essays. He had all the information he needed to write a complete essay by himself and didn't want to work with the other students. I told him he had probably better contribute to the group work because the teacher had structured the class that way, but that that shouldn't prevent him from writing his own essay apart from class. So he did. The self-motivated essays by these students are included in the Samples section of this book. Essays by fellow students are often as inspiring as those by professional scientists.

I will briefly recapitulate some of the approaches normally successful for finding viable topics: 1) the students may refer to the writing about themselves and follow an interesting line of development from that material; 2) they can discuss finding a topic with classmates, who will help guide them to fruitful ideas; 3) a student or the teacher can lead them in a brainstorming session; 4) they can peruse magazines, newspapers, and books, including textbooks, until they are inspired by an idea; and 5) you can present yourself as a resource, and guide them to specific subjects.

Please take the time to design your own strategies here, and to summarize the thoughts and questions you have had along the way.

(Pause.)

TOPIC EVALUATION

Essays turn on ideas. Ideas are very energetic entities. With the right idea, a student can feel as though the essay is writing itself. (One of my ninth-grade students once said, "I have all my ideas when I stop trying to think.") A topic is not an idea, it is only a subject, inert mass, potential energy. Before writing about it, the student needs to figure out how rich with ideas it is, and how strong the ideas are. To be strong, the ideas must be interesting to the students themselves. To consider writing to be of value because it is interesting is a novel experience for many students. As a result, they may have difficulty believing that they should be honest in their assessments. We should do everything in our power to convince them to be honest, that they will waste everyone's time if they are not.

It is impossible to predict the potential of every topic by evaluating it in advance, but the most problematical can be weeded out easily. Topics can be rejected later, for unforeseen reasons, but it is not a good expenditure of energy for a student to spend a lot of time writing about and researching a fruitless subject.

The process of evaluation and the generation of new ideas are inseparable in the development of an essay, and one grows organically from the other. For the sake of discussion, however, let's imagine that determining the strength of a topic takes four steps.

The first step, and perhaps the most critical, is to have the students do directed freewriting in response to the question "How and why is this topic interesting to me?" Topics that had seemed extremely provocative at first may prove dull when inspected under this lens. Conversely, a seemingly barren topic may in fact prove rich with interesting ideas.

Here are some examples of directed freewriting on the interest of a topic:

> My topic is the modernization of computer technology. I believe that by the time I graduate from or am in the midst of college, computers will have dominated the world of technology.
>
> My feeling about modern computer technology is that computers are great. They stand for or represent an intellectual standard in society. However, if they become advanced any further, we the working people will have no purpose in society. Will we have jobs? If we do, what kind of jobs will they be? Will we be forced to work in fields of no interest to us because of computer technology? I am really concerned about this. I do not know if any of these questions can be answered. Maybe only time will tell.
>
> For example a specific article in the *Times* talked about how computers can help students with algebra. —*Shawnel Boone, twelfth grade*

> My topic is hypnosis. I don't have a good background on this topic but I have been interested in it for quite a while now. Based on my little knowledge of this topic, my definition of hypnosis is that it is a state of being where someone else is in control of what you do or what you say. Hypnosis is a way of searching into a person's mind to acquire information that they have blocked out of their conscious mind. —*Carmen Santiago, twelfth grade*

Why write about sea cows? Well, why not! Although I admit that it was something of a random selection, in no way does that minimize the importance of them.

Here is an animal with feelings second to no other species in existence. As a matter of fact, no living animal should be categorized or ranked. Which we do even among ourselves. No one really fully understands the total potential of ourselves, much less those of other animals. Maybe this should never be found out, then again, maybe it should. Only time will give us the answer.

So as I begin to tell you about them, don't laugh or get bored, but listen and learn about something that we may learn about and also something that may be vitally important to us in the future. You just don't know how things may turn out.   —*Edgar Lantigua, eleventh grade*

The second step, which begins the process of developing the essay itself, is to write all the questions that come to mind about a topic. The student can benefit from this by doing it individually. If there is time, and the classroom situation seems to justify it, having others contribute their questions can also help the writer. The questions may immediately lead to observations, and to further questions. (For an extraordinary example of questioning, where the process turned into an essay in itself, please see Diane Spann's essay "Television in Society" in the Samples section.)

The third step is to spend a period of time trying to find a slant on the topic that will narrow it to manageable proportions in a natural way. Picture the slant as a magnet that is to be propelled through a circle containing iron filings: information. The force propelling the magnet is the writer's intuition. Intuition plays a pivotal role in the development of any authentic piece of writing, whether in the arts or in the sciences. A writer will *feel like* or somehow *know* an approach is going to work out. Some characteristics of the feeling are optimism, inspiration, and a willingness to take intellectual risks.

A writer's slant may start from different points on the circle from, for example, using questions, as in step two above. The question may or may not prove to be appropriate. In one of my classes a student had chosen the topic of world hunger. There were plenty of good questions to ask about the subject. He ultimately chose this one: "Is there a difference in the hunger of an African tribe that has not eaten in days and a city dweller who is reduced to handouts or starvation?" While the question was certainly thought provoking, it didn't prove to be appropriate. The student said, "In my research I have found that there is no essential difference," and concluded, "The real problem is that we're too selfish to work together." The sentiments were certainly real, but there was not enough sustained interest in the question. The slant proved weak.

Another student, whose topic was the future (which looked at first blush like a subject bound for a bog of generalities) posed this question: "Will life survive?" Surprisingly, it became a good start, leading to other questions: "Will *human* life survive? If it doesn't, will it be exterminated by war? Disease? What can be done about it?" The questions led to a nicely balanced meditation on two different types of thought concerning the future: optimism, which is characterized by the belief that technology will solve the problems that it creates, and pessimism, which is characterized by the belief that technology will compound the problems.

If the slant doesn't come from posing the right question, it may turn up in research, or from the drift of material in a first draft, or from the form the writer chooses (book report, speech, magazine article, meditation, etc.), or from a consideration of the audience that will hear or read the essay. Whatever brings it about, it is essential that the student's intuition be allowed to function. Neither originality, energy, nor commitment can exist without intuition. The magnet will lose its force and stop in the circle.

We may have to exercize a great deal of patience as students try to find good slants. In *The Open Mind*, J. Robert Oppenheimer used a different analogy to describe the process. "The experience of science—to stub your toe hard and then notice that it was really a rock on which you stubbed it —this experience is something that is hard to communicate by popularization, by education, or by talk. It is almost as hard to tell a man what it is like to find out something new about the world as it is to describe a mystical experience to a chap who has never had any hint of such an experience." We may have to slow our expectations, without lowering them, and be silent while the students stub their toes.

The fourth step in evaluating a topic is discussion. Depending on the size and format of the class, students can discuss their directed freewriting, their questions, and their proposed slants in pairs, small groups, or the whole class. In the discussion, the central questions to be addressed to the writer are: "Why do you want to write about your topic that way? What will it show us?"

The results of discussing these questions can be surprising. Despite praise for their topics, students might still decide to drop them. Conversely, criticism might only strengthen their resolve to write about them. Moreover, the questions may not need to be answered at all to help a student. It may be enough for the writer to feel that the other students are saying to themselves, "Oh, I've always *thought* about that topic, but never talked about it, and I'm dying to see what you have to say about it."

To review briefly: Topic evaluation and the generation of ideas are processes that blend together, but which may be described in four steps: 1) directed freewriting on the question "How and why is this topic interesting to me?" 2) directed freewriting that includes all the questions that come to mind about the topic 3) searching for a slant on the topic that feels intuitively right, and 4) discussing the topic with others.

Please pause to write your thoughts, and perhaps devise a way to include the topic evaluation as a tactic in your strategy to develop good essays in your classroom.

(Pause.)

If your students are straying uncomfortably far from the subject of the course you teach, perhaps you would like to see if they can establish a connection between the subject and their essays by having them do a short piece of freewriting answering this question: "How does (the subject of the course, e.g. General Physics) affect the topic of my essay?" If the results show that the subject of the course can be woven in meaningfully, then the student has already begun to do so.

RESEARCH

While many slips between the cup of information and the lip of students' notebooks may occur during research, it is one of the most stimulating phases in the writing of an essay. As a teacher, you know your own situation best. You know where sources of information are, how much you can expect your students to do outside of class, how effective the library is at channelling students to relevant books and magazines. You may have your own collections of source material that you are willing to share with the students. But whatever your circumstances may be, there is an enormous potential here for students to make discoveries on their own.

The key, as in previous phases, is to allow the students to understand and solve their own problems as much as possible. A few minutes of directed freewriting at the beginning may help them to develop their research plans. If they carry out their plans successfully, or find alternatives quickly, they will need little supervision. If they run into problems and need suggestions, they might find it interesting to write a summary of what happened, before talking to you or getting help from the librarian or a fellow student. If they are goofing off, your best response may be to turn them back to the original motivation for the whole essay: their interest in it. Have them analyze why their interest is waning.

When I asked students to find information on their own time outside of class, only ten or twenty per cent of the students came through. (I was meeting with them once a week.) This didn't mean the assignment hadn't been worthwhile—certainly for those students who did it, it was valuable. Realistically, however, to give the students a chance to experience the rewards of research, I had to commit class time to the effort. I took them to the library myself and, with the help of the librarian, let them pursue leads. I kept regular tabs on each student's progress. Here are a few of my diary entries:

> Kenny: Came and asked what to write about himself. We discussed what his problem was with the topic of terrorism: though he is fascinated with it, the research involved is too awesome (he feels as though he has to tell the history of the entire PLO struggle with Israel) and the generalizations he is able to write are quickly done. So he finishes too fast, before his interest is used up. I suggested writing a short story. Now he is writing about whether that is a good idea while simultaneously developing the idea. He tried.

> Tor: Stared at large chart book the whole time while talking to a friend. Then, at my goading, got another book to look at. Coasted.

> Louis: Read a psychology book the whole time. I advised him to jot down notes. Tried. Check next time.

> Edgar: Writing something he's going to keep mysterious for a while. Tried.

The success rate of this method was one hundred per cent if you take into account that results did not necessarily appear in the form of information in the final drafts of essays. Often, the lesson was *not* to use information. Kenny, the student researching terrorism, decided not to write a short story but to go ahead and pursue research, though not about the PLO-Israel conflict. Unfortunately he picked an amazingly turgid tract from an academic text about Bolshevik terrorist activity, and instead of telling me about it, tried to understand it and include it in his essay. He was later relieved to discover that because interest—his, mine, and the other students'—was the main criterion, he was better off cutting most of it. We are reminded that research material needs to meet the same requirement as the essay itself: is it interesting?

Once students find good sources of information, they can benefit from being aware of many things: questions being answered, authors' slants, intended audiences, major points of interest, good quotations, relevant statistics, new vocabulary and phrasing, graphic techniques, etc. — but the most important thing is their own reaction. Either concurrent with their reading or in their summary writing at the end of the research session, they should note what they found most interesting, how they might use the material, and their questions about it. Among these notes they should include their feelings. If they are using dialectical notebooks (a technique in which facts are noted on one page, personal reactions on the opposite page), they know to record their feelings on the side opposite their notes. If not, they may simply benefit from jotting them down anywhere. These may be the most important notes they take. Lest we still believe that intuition in science is not to be trusted because it is too subjective, let us look at a quotation from Harvey J. Gold's *Mathematical Modeling of Biological Systems*:

> The results of a mathematical development should be continuously checked against one's own intuition about what constitutes reasonable biological behavior. When such a check reveals disagreement, then the following possibilities must be considered:
> a. A mistake has been made in the formal mathematical development;
> b. The starting assumptions are incorrect and/or constitute a too drastic oversimplification;
> c. One's own intuition about the biological field is inadequately developed;
> d. A penetrating new principle has been discovered.

The point is not that one's feelings are going to be a measure of the *truth* of what one is reading or observing, but a measure of one's *involvement*. These notes during research (or experimentation) could well become the meat of the essay.

Many students approach research reluctantly because in the past they have had to do it mechanically, without regard to their own interest in the information. For this reason they shouldn't be blamed if they hesitate, or try to avoid it. Ideally, we will have time to guide them to captivating information and leave them alone long enough to enjoy it. If we don't have the time, some students will find such information anyway, but some will not, so we will not be completely effective in changing their attitudes. Even with limitations, however, we can disarm the attitude by insisting they be honest in their essays and discard inappropriate information. And we can reinforce their own sense of themselves as young scientists by showing them, through grades as well as personal responses, that it is the vigor of their curiosity that matters most.

Please pause here to collect your thoughts and to write an assessment of your own views of student research.

(Pause.)

ROUGH DRAFT

Imagine all the writing the students have done up to now, with the exception of the freewriting, to be different metals, and that we want to melt them together to create a strong alloy. The rough draft is simply the combining of the metals into the pot. The melting will come in the heat of peer criticism and the alloying will come in the cooling of revision. It is enough now simply to have the students put together all the material they find interesting in as smooth a way as possible. If they doubt the value of some of the material they have picked up, they should include it at this stage anyway.

Before proceeding, however, it might be profitable to stop and look one last time for links between their topics and the subjects of our courses. In one class, I asked the students to do ten minutes of directed freewriting on the subject and, after five minutes, saw they were writing avidly, so I asked them if they wanted the period extended by five, to a total of fifteen minutes. Half the class said they did, so I extended it. Here is an excerpt from Kenny's directed freewriting. You will remember that his topic was terrorism:

> In one way I don't think electronics has anything to do with what I'm working on. My essay is on terrorism, and I don't see what that has to do with electronics wait, I just thought of something. One way that I think electronics and terrorism are related is. . . the U.S. may know where they are in Iran by computers.

While all of the students found links, they didn't necessarily use them in the final revisions of their essays, but the directed freewriting definitely fell into the category of a "microtheme." It also started the job of writing the rough drafts for many students, making the beginning less daunting.

In my situation I knew the students would be as articulate as possible, and that problems of clarity and detail would be ironed out later. It did seem necessary, however, to talk about conclusions. To make the next phase of the process, peer criticism, work well, those students whose essays required conclusions needed to have something, no matter how sketchy, ready to discuss, or there wouldn't be enough to talk about. I pointed out how conclusions worked. I said, "Give a brief summary of the points made in the essay, lining them up as forcefully as possible. Introduce any new points or observations you would like to make. Recapitulate the opening statement. (I defined 'recapitulate.') State your opinion as forcefully as possible." With that exhortation, I left it up to the students to combine their metals as they wished.

If your goal is to get the students to write in a certain format (a speech, a book report, a videotape script), you should remind them what that format is, but otherwise perhaps you should let them alone. Interesting forms and structures will emerge in their writing because you have inspired them with good examples, because they have noticed different ways of approaching topics in their research, and because they have chosen strong slants. It may not be so easy to leave them alone, though. Like me, you may have been taught that essays should always be written in five paragraphs, or that they have to have good introductions, expositions, developments, themes, and conclusions, or that they have to present pro and con arguments. If our traditional training can remove an obstacle for a student, fine; but if it cannot remove obstacles, it is liable to create them. Why impose structures on the students when they don't need them?

The one critical problem that cropped up in my students' rough drafts was a loss of confidence. At the last minute, some stopped believing their writing was interesting and retreated to the safe haven of information, out of the risky world of ideas. They retreated to the porch. The only trouble was, there was no house attached to that particular porch. If they had retreated to themselves, they would have had better protection. Happily they regained confidence during peer criticism, when other students reassured them that what was *authentically* interesting was their own ideas.

Please pause here to write your thoughts about rough drafts.

(Pause.)

PEER CRITICISM

Our efforts to encourage students to respect their ideas will pay off most eloquently in this phase of essay writing. They are excellent critics of each other's work and will apply peer pressure to see that every student's writing is as good as it can be. As teachers, all we have to do is facilitate the process. Students may criticize each other's work one-on-one or with the whole class. Each class will require a different mix of the two methods. Whatever the mix, certain tools will be useful.

The first is a dignified atmosphere, which will affirm their mutual respect. Approaching the process with solemn formality at the beginning helps to create such an atmosphere. While many aspects of essay writing (as we are defining it) may be loosely structured, this is not to be confused with lack of formality. In peer criticism we are handing over our roles as teachers to the students, and teaching being a principled activity, the students need to appreciate the same principles. To do so I always make sure the words *peer* and *criticism* (in a broad, positive, literary sense) are defined, then I point out that the success of the writing will depend on the seriousness and effort they put into the analyses of each others' work. If a student fails to do a good job, it is not his paper that will suffer, but that of a classmate. They are responsible for each other. While I emphasize this, I also provide a safety net and tell them that if they feel they have not received thorough criticism from the classmate who worked on their paper, they can bring it to me and I will either criticize it for them or redistribute papers until they are satisfied.

The next tool to give them is the language of criticism. Having kept myself abreast of the students' work in one class, I deduced that it would be appropriate to hand out a list of definitions of these words: *analogy, anecdote, behavior, category, conclusion, correlation, description, intention, method, objective, observation, premise, research, speculation,* and *theory*. We didn't discuss all these words, but we did review the meanings of the most pertinent, which proved to be: *conclusion, description, example, hypothesis, inference, intention, premise, speculation,* and *theory*. Perhaps the most important was *intention*, since the primary responsibility of a student criticizing another's paper is to see that the writer's intentions are realized as fully as possible.

To criticize well, one must address the right questions. I found it helpful to hand out the following list in situations where there was ample time to have an open discussion to show how the questions should be used. Without such a discussion, students are liable to treat the questions like a test. In any case, the questions might be a base for you to create your own, in case you haven't already done so. Please note that these are only *some* of the questions to consider when criticizing an essay:

1. Is the author's premise or hypothesis clear?
2. If not, why not?
3. If there is no premise or hypothesis, what is the author's intention?
4. Are the arguments and assertions sound?
5. Are they supported by the facts?
6. Do the facts originate from good sources?
7. Has the author attributed the sources adequately?
8. Does the essay contain the author's opinions?
9. Should the essay contain the author's opinions?
10. If so, are the author's opinions justified?
11. Is the writing clear and free of contradictions?
12. Is there enough detail? Too much detail? Where? How could the problem be corrected?

13. Is the conclusion logical?

14. Is the essay informative and enjoyable to read? Why? Why not?

Using this list, one student wrote the following criticism of a classmate's paper:

> The author's premise is clear. However, towards the end of the essay he goes off on a tangent. His opinions are valid, but I feel that they should be more concise and better incorporated into the essay.
>
> The author's intention is clear, but he contradicts his premise with his hypothesis. He first states that it seems as if the school will not remain as is (in a few years), then he says it will. I think his essay would be stronger if he takes a stand on one of these two opinions.
>
> The arguments and assertions are sound. They are supported factually. However, he should incorporate a few more facts and statistics to strengthen his essay. He should get a few more sources to support his opinion.

My only criticism is that the critic did not tell the writer if the essay was enjoyable to read. He could have used a bit more praise.

The next tool to give the students is time. To analyze a paper may take an entire period or longer. In the most pressing circumstances, I have had students do it in as few as twenty minutes, but the results were not thorough. If you are having the students swap papers, or are gathering them and handing them out on some other individual basis, ideally there will be time to 1) do the criticism, 2) analyze with the whole class a few examples of the criticism to show how, for instance, specific observations and examples need to be made, 3) redo the criticism and allow the criticizers to discuss matters with the criticized. If you are reading each paper to the whole class one after the other, then you must have all the time you need, and you are very lucky.

In either case, some discussion should take place, to fan the heat under the students' ideas. Open discussions in class can show students exactly how effective they can be at helping shape each others' work. In one of my classes, a student was advised to research more than one case to make a point about surrogate motherhood. Another student had done research but hadn't reacted to it thoroughly enough. Redundancies were exposed in yet another piece. Criticizing the writing openly, in a respectful, dignified atmosphere reinforced the students' sense that they weren't alone in their writing, that their ideas mattered to others.

It is by peer criticism that students who have resorted to the thoughtless regurgitation of information will fully realize the uselessness of their writing. In one class that was particularly oriented toward regurgitation, I found that there were several distinct symptoms of students' lack of concern: contradictions, inadequate summaries, unsupported assertions, and unintentional lack of clarity. I selected the following examples from their work so the students could see what happened when they weren't writing about what really interested them:

Contradictions: "Organ donation does not cost too much but it is a little expensive." Or: "The budget of the Militarization Space project is very limited due to the fact that a lot of the budget given for military use is going to be based on this project."

Inadequate summaries: "We can conclude by saying that cruelty to animals is certainly unjustified, but how else can we advance in fields such as medicine without the aid of animals? Evidence proves that animals in the past have been treated extremely cruelly."

Unsupported assertions: "The biological mother feels that she should do it [become a surrogate mother] as a favor toward those who are unfortunate that they can't have any children."

Unintentional lack of clarity: "How could someone become a surrogate mother? New York would be the first to accept surrogate parenting. It is organized by the Deputy Senate Majority Leader John Toohue. This event is a relationship with the Baby M custody case in N.J."

These cases are examples of laziness on the part of students who are capable of high motivation and clear writing, but who simply haven't cared enough about it. Another student was criticized by his peers for *intentional* lack of clarity. He advocated greater expenditure of funds for the Star Wars system because the Soviets were building the "Red Shield," but could not quote any statistics about either project, nor about funds that other students were concerned would be depleted: welfare and social security. He said he didn't think the use of such statistical detail was necessary. Students accused him of being obfuscatory to protect his political bias (he had declared provocatively at the beginning of the discussion that he was a "solid Republican"). They said that in order to persuade people he should use detail. Backing away from his assertions, he said he wasn't writing to persuade, really. But his peers *wanted* to be persuaded. They wanted to hear the statistics. They *cared* about his position, and were disappointed that his essay didn't provide a good basis for discussion.

Please stop here to record your thoughts about peer criticism.

(Pause.)

REVISION

Before revising their essays, students need to assimilate what they discovered in peer criticism. A good way to do this is to have a period, perhaps fifteen minutes, of directed freewriting on the issue. In this piece of reflection, they would do well to remember all the criticism they have received and the feelings they have about it. They should add criticisms of their own, as of another's paper. They should cull observations they find to be valid from the ones they disagree with. They should note any new ideas or inspirations that come up. And they should form a revision plan, incorporating any new research necessary.

By reading this directed freewriting we will understand the students' needs and be able to meet them by adjusting future lesson plans accordingly. Naturally it would be best for the essays if we could commit all the class time necessary to take care of every need, but even if we can't, the students will have benefitted from the process.

The amount of revision we can expect will vary greatly from individual to individual, depending on how much interest they still have in their subjects. Some may not revise at all and others may write several more drafts.

After students have finished their revisions, they should write summaries of what they accomplished, and, if they didn't accomplish all they wished, the reasons they didn't. We should read these summaries, because they will be helpful in interpreting the experience of the students and in planning future workshops, labs, and classes. I had my students do this summary writing during evaluations at the end of my workshops. (I will quote from examples later.)

While we are likely to be satisfied with what most students ultimately decide to include in their final drafts, some may disappoint us. One of my students, who was writing about teenage suicide, had written some very interesting directed freewriting on the subject of the attempted suicide of former national security advisor Robert McFarlane, trying to define its effect on young people, but he didn't use the writing in his essay, even after I suggested he do so. He said he thought it was too complicated to get into the politics. Here are my notes after our conversation: "Is this a failure of mine? If they don't want to write, they don't have to, and this is the consequence—but I'll say this, François' paper, when he read it, was interesting. *He* was happy with it, glad he had done it, glad to be part of the class, and to have gone through the thinking process." Still, I wish he had put that material in.

Another thing that bothered me was the non-scientific nature of some of the essays. For instance, one on astrology was entertaining but lacked systematic objectivity. Somehow it got through peer criticism making unreconstructed assertions such as "People born within the same sign see life in the same perspective and their wants and needs are similar." I included the essay in our student anthology nonetheless, because I had maintained

33

all along that students would determine the validity of their own writing, based on their interest in it. A student editor, striving for high scientific standards in an anthology, might easily have rejected it.

Though the other students' essays were more scientific, a low percentage of them ultimately chose to write about the subjects of their science courses except in their "microtheme" exercises. Ideally they would have been so caught up in their subjects that it would have been natural to write a great deal about them, but when curriculums have been reduced to dry information, it is only natural that students avoid them in self-motivated essays. My students were as alienated from writing as they were from science, however, and by learning to respect their own ideas first, they prepared themselves to approach the subjects of their courses with more interest.

Please pause again to write your responses to the ideas in this chapter. In particular you might determine how much time and effort you can commit to the important process of revision.

(Pause.)

PUBLICATION

Publishing the finished essays in an anthology is worthwhile for a multitude of reasons. Besides giving a tremendous boost to the students' self-confidence, it is an affirmation to the teachers, administrators, and parents who have put effort into the project. It is a climax, a culmination, and a cause for celebration. Anthologies also serve as foundations for traditions in a school. Subsequent students can read them and emulate them. Word gets passed down from year to year that the writing program is a cool thing to do, and teachers have an easier time using the techniques and procedures.

One of the reasons anthologies are valued is the amount of time, money, and labor it takes to produce them. To arrange for all three takes good planning, which will need to be done before the writing is even begun. Please pause here to assess resources available to you for the production of a student anthology. You may want to consider the following questions: What funds are available? How can they be acquired? Is there a department in the school already equipped to print an anthology? Who will comprise the staff of the anthology? Students? Parents? Teaching assistants? Myself? Does the potential staff have the appropriate skills: editing, typing, printing, binding, and distribution? How wide should the distribution be? How many extra copies should be printed to pass out in next year's classes?

(Pause.)

(Also, refer to the question "How does publication enhance knowledge?" in the Questions and Answer section for more thoughts on the subject.)

Publishing an anthology of the students' essays is desirable for the reasons I have mentioned. Having the students read them aloud is more than desirable, it is essential. The class itself is sufficient for an audience. Reading completes the process of writing, in a visceral and immediate way that cannot be achieved by any other medium. Some students will resist reading their papers aloud. Most of these can be persuaded after they see the good nature of the responses to other students reading. Often work is applauded. Always, because of the processes that have come before, the work is respected. If a student continues to resist, we or someone else in the class ought to read his paper for him. All of the papers should be read, because each student needs to feel the effect of his words on others. Readings can be videotaped for viewing by future classes as well. If used with discretion, they can establish traditions within a school as effectively as an anthology.

Please pause again to jot down further thoughts about anthologies and new thoughts about readings.

(Pause.)

EVALUATION

Having the students evaluate their experience of writing is instructive for us, but it is also the final confirmation that the process has been for the students' own good. It is as much a way for them to say thank you as it is an opportunity to help improve the situation for those who follow. Please pause here to write several questions that you might include in a student evaluation.

(Pause.)

Here are the questions I gave one class:
1. What were your expectations when this work began?
2. What did you think of the prospects of writing an essay?

3. Did your thoughts about writing change during the course of the workshop?

4. What did you find most interesting? Least interesting?

5. Do you feel that writing this type of essay is appropriate for an electronics class? Why or why not?

6. If you had it all to do over, what would you do differently in writing your essay?

7. How could the writing workshop be improved?

Except for a couple of extremely alienated students, all in my particular workshops found writing science essays interesting. Some thought it was going to be boring at first, but "as we started giving our points of view and discussed different aspects of science, it started to get interesting." They liked the fact that they could express themselves "freely and openly."

Besides finding it interesting, many thought it greatly improved their writing skills. Some, while not exactly falling in love with writing, still thought it was useful because, for instance, "It gave me tips to write better papers in college."

The biggest argument about the workshop was whether it was relevant to classes with subjects like electronics or physics. One thought it was "because it made me look at science as a whole in a different perspective. I was particularly impressed with the methods that Mr. Worsley used to make physics more romantic and interesting." The same student thought it would be a good workshop for any subject.

Many felt ambivalent about its relevance. I'll let one student speak for several: "Although it is true that all subjects are interrelated, I disagree with the introduction of this program in a physics class. It is true that writing about the sciences in a creative manner will create a better ability to understand and learn, but it can possibly confuse those who will in the future have to write compact, non-opinionated science reports. I think to improve it we should have a longer time period in which to work and in this time period we should be guided into writing a traditional science essay."

The other point of view expressed with some frequency was that while it was interesting, such writing should be moved entirely to English classes. Beyond the arguments implicit in this book which, while they don't preclude enthusiasm for writing science essays in English classes, do argue for its value in science classes, there is another point I would like to add here. English classes are often geared toward the subjects of American and English literature, not to writing. Furthermore, English teachers have not necessarily been trained to teach writing and, as a result, feel as inadequate, at first, as science teachers. I hope the techniques and ideas in this book will help remedy that inadequacy in both departments.

Here are some other student comments, quoted at random:

> The poetry class was fun and the freewriting was relaxing.

> It demonstrates how important it is to put your words or theories into words. A lot of people have great ideas and questions, but do not know how to express them in words [to make others] understand.

> It helped develop a writing skill rarely taught, the usage of your opinion.

> I became more in-depth in my essays for other classes.

> The thing I thought was most interesting was the ability to criticize quietly, independently, and freely.

> The students didn't give enough output. It could've been better than this.

> The first couple of weeks I really didn't like this class. On occasion I was going to cut. But I didn't and it's a good thing, because I really like it. Now I enjoy coming to write essays, and discussing them.

The last quote was written by Kenny, the student who, in his essay about terrorism, battled with the terrors of research and won.

Please pause again to record your evaluation of this presentation of methods to write an essay. It is appropriate that the last word be yours.

Writing Experiments

Misunderstanding

Have your students write about things they used to misunderstand when they were much younger, especially things related to science. They could write in prose paragraphs or in free verse. Each line of the poem could begin with "I used to think. . ." (a simple writing idea described by Kenneth Koch in his *Wishes, Lies, & Dreams*) or "I misunderstood. . .," as in Koch's own poem "Taking a Walk with You":

> My misunderstandings were:
> 1) thinking Pinocchio could really change from a puppet into a real boy, and back again!
> 2) thinking it depended on whether he was good or bad!
> 3) identifying him with myself!
> 4) and therefore every time I was bad being afraid I would turn into wood. . .

You could also assemble lines from all the students and read them aloud as one big collaborative poem.

This assignment often produces entertaining and amusing works that can give students insight not only into what they know, but *how* they know it.

Invisible writing

Invisible writing can now be done most easily on a computer by turning off or darkening the monitor. Not being able to see what you're writing changes the way you write it. For one thing, your memory of what you've just written often becomes more visual and imagistic, as does what you now write. The writing produced in this way can either be made visible or permitted to disappear. Invisible writing is like Alice in Wonderland: how can I know what I think till I see what I say?

Number tales

Numbers make wonderful characters for a story or a poem. They all look different, just like people, and they all seem to have characteristics that belong only to them. Furthermore, you can combine them in all kinds of ways to make things happen in your story. Or you can simply list a cast of characters. Here are some samples.

A 2 is a pregnant 1. This is obvious, not only because the 2 has such a big belly, but because everyone knows that to get 2, you must have 1 plus 1.

The number 10 is a marriage of here with nowhere. The number 1 is all here; it says, I am and I do. The number 0 is empty. It says, I am here to hold. When they stand side by side to be married they make 10. Under 10 are found all the numerals that there are.

An 8 is its own mirror. Backwards, forwards, upside down, it is always an 8. It is also a headless snowman.

An 8 is the first cube. $2 \times 2 \times 2$ makes 8. With these cubes, you can build the castle of 8's. And let three 1's come live there. But $1 \times 1 \times 1$ is still only 1, because 1 casts no shadow.

A 3 is a pair of lips that pronounce the word "three."

Loons and whales

Turn the classroom into a different place, in this case a lake or an ocean. Get tapes and records of the communicating sounds of loons and whales, easily available from libraries, record stores, or natural history museums. Write while listening to the sounds, preparations for writing might include: a discussion of the sounds and their possible intentions; eyes closed while listening; lights off while writing; stopping, discussing, and writing again.
Some examples:

These sounds are very strange, as if they are calling someone or something. The flute is what they sound like. One of them sounds as if they are calling out my name. It is such a bunch of strange and interesting sounds. Some of the sounds are as if someone is scratching on a door, another sounds like a dog barking at something. But there is one sound that is repeated constantly. . . . I can hear the wind blowing, like a harsh blow. Then too it sounds like an airplane. The shaking of the trees as the wind blows or the plane goes by. That one sound continues and it won't stop. Is it crying? Or some sort of signal? . . .
—*Kolie Marcus*

JOHN THE WHALE: Hi, Susan. How's your mother doing?

SUSAN THE WHALE: Aw, she's alright. She had to go see Dr. Blue Whale this morning for a checkup.

JOHN: Did you see my brother Tommy today?

SUSAN: Yeah, I saw him. He was by the coast. I warned him about that but he didn't listen.

JOHN: Well, I'm just gonna have to tell my mother about that, he knows it's dangerous. He could get caught and you know how them stupid humans are?! They're just plain crazy! All they want to do with us is hunt us, eat us, and melt our blubber for oil. Them no good humans should perish from the face of this earth!

SUSAN: You're right! But I have to go now. I have to get my fins done, they must look awful!

JOHN: Bye Susan.

SUSAN: Bye John.

—*Rodney Pink*

An ensuing discussion could include the subject of communication with "people" from other planets.

A mock test

The following test is a good one for students who are not accustomed to being allowed to think for themselves and write what they like. It could be adapted to the circumstances of different schools by altering the questions, or by using it as a model for entirely different questions. Suggested instructions to the students: "Take the entire period to answer at least five of these questions thoroughly. If you want to answer more, please do so. If you want to answer fewer, that's all right too." The instructions contradict themselves, but that is precisely the point. The *forms* of testing are being followed, but the content is altogether different.

TEST

1. Do you believe extra-terrestrial life exists? If so, what should we tell it about ourselves?
2. How do you imagine life got started on earth?
3. How did humans evolve from earlier forms of life?
4. How would you change the school system to serve you, your culture, and your society better?
5. What do grades have to do with learning? Do you learn better because of them?
6. What is your opinion of the light in this room?

7. What would you rather be doing now, instead of answering these questions?
8. Under what circumstances do you really feel like you're getting somewhere in your life?
9. Should students be tested for drug use in the schools?
10. Why do some people like drugs? What happens to them chemically?
11. Should athletes be subjected to mandatory drug testing?
12. Why are there more teen-age pregnancies in underprivileged neighborhoods, and what are the consequences to society?
13. What is the "military-industrial complex" and how do you fit into it as a student who may go into one of the armed forces' education programs?
14. If humans can control their future evolution, what should they change about themselves?
15. Who will control how life is changed genetically in the future, and what motives will they have for determining how it will be changed?
16. What do you think our chances are of getting involved in a nuclear war? How do you think it will begin, if one occurs?
17. Do you enjoy learning science? Explain.
18. How does science affect your everyday life?
19. How does science affect your imagination?
20. Do prejudice and racism still exist in our society? If so, are you affected by it?

Being scientific as a writer

Aside from having science as one's subject, one can incorporate scientific ideas into the construction of any piece of writing. For instance, make different sets of index cards for the categories: information, interpretation, new ideas, hypotheses, etc., about a particular subject to be written about, such as clouds. Each card should have a title (cloud shapes, cloud literature, etc.). Tape them to the wall before your desk or table, so you can perceive the whole at once. Then make a chart or diagram that gives a visual representation of how aspects of the subject relate to each other. Include oblique or tangential material. Next create a verbal structure or outline for the piece. Tape the diagram and the outline to the wall so that a brief glance at all the material can remind you of where you are or where you might want to be. This step can be followed by a brief free associative writing on the subject, which will provide relaxation and focus. Then proceed to write.

The writer might also enjoy putting different paragraphs or sections of one piece of writing on different sheets of paper, then spreading them out on the floor and rearranging them in the way that seems most—and then least—logical. Draw conclusions from each. Another possibility would be to make a random arrangement of paragraphs or sections to see what you come up with.

Instead of writing in essay form, it is unexplainably illuminating to create a set or series of reflections on a subject and number them sequentially, chronologically, or randomly.

Dialectical or double-entry notebooks and collaboration

Critical and multi-faceted thinking between and among students. Fold the facing pages of a notebook in half, creating four columns (see fig. 1). Use one column for taking notes on a text or idea. Use direct quotes, with citations. Use another column to make comments on the text. Give the notebook to someone else for comments on the text and the original marginal notes. This process can continue indefinitely, through a series of readers or back and forth.

Other methods of collaboration are:

- Groups of two to seven students working on a problem together in the classroom. One student can serve as notetaker; the teacher joins one of the groups.
- Homework that involves people working together to solve particular problems.
- Writing about a problem collaboratively, for example in alternating sentences.
- Team teaching: three teachers working together in a class, one as the "teacher," one as the assistant teacher, one as notetaker. Or, teachers arguing a concept together before the class.
- Students taking over and teaching the class, designing new writing experiments.

Dream writing for problem solving

Attempting to dream about the solutions to mathematical and scientific problems and then to write about them. Have students concentrate hard on a specific idea before going to sleep. It's good for the whole class to focus on the same idea. When the students wake up, they should write

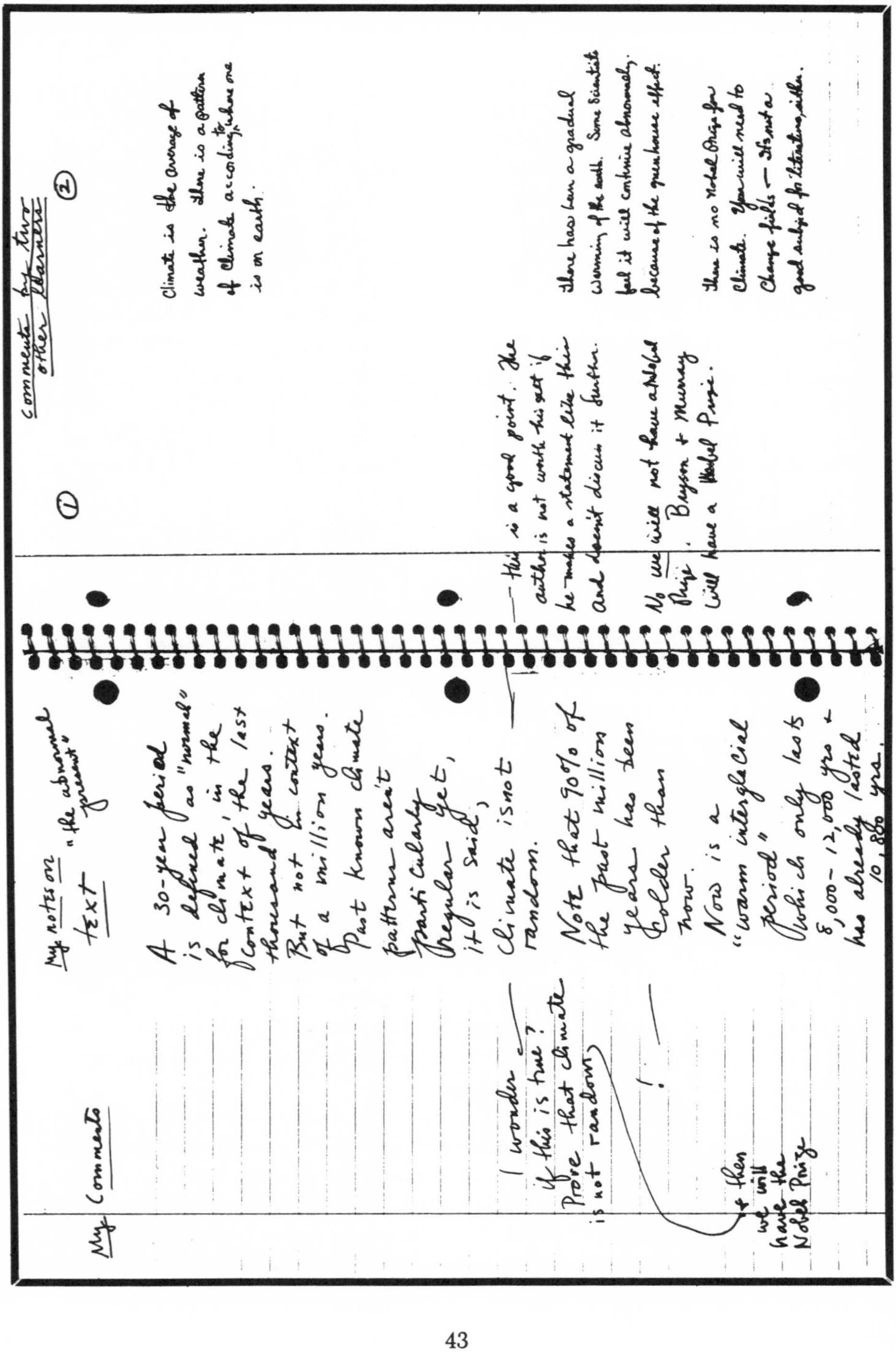

Fig. 1: Double-entry notebook.

down their dreams (or tape them). Then discuss them in class. The dreams can be turned into narratives, essays, equations, charts, diagrams, science fiction stories, or any form of writing that seems to reflect the dream, including the entirely speculative.

Students and teachers often have interesting dreams about science without noticing or cultivating it. Dreams about mathematical and logical questions can be especially amusing.

For those who have difficulty remembering dreams, here are four suggestions: when about to fall asleep, rest your elbow on the bed and raise your hand and forearm and you will find that when your arm falls down you may be able to retrieve a dream; read a book about the subject to be dreamed about right before going to sleep and keep the book next to you; have someone awaken you after two or three hours of sleep (always keep paper and pencil within arm's reach); arrange to be able to wake up in the morning and then go back to sleep for a while (dreams had at these times are easier to remember, as are nap dreams).

While writing this book, I had two dreams relevant to this experiment. I dreamed that A, B, C, D, and E had to be set in a vertical column and had to be next to an unknown column. The alphabetical column had to remain forever the same, the unknown column always had to differ. The dream showed me both a sample series (the one on the left below) and a series that was "correct for today" (the one on the right below). So this was what I saw:

A	x^{a-1}	$A = x^{a-1}$	
B	x^{b-2}	$B = x^{a-2}$	
C	x^{c-3}	$C = x^{a-3}$	
D	x^{d-4}	$D = x^{a-4}$	
E	x^{e-5}	$E = x^{a-5}$	

I didn't know if an equal sign really belonged in the second column or not, but I saw quite clearly the colors of both the numbers and letters, and the numbers began to take on the colors of the letters, so that though 5 began as tan, both E and 5 were definitely green, though "a" was red. Merry Christmas! I realized I had been reading Rimbaud's poem "Vowels," in which each of the vowels is given attributes of color combined with metaphor, but wherefore these dream "equations," or were they just juxtapositions?

A friend of mine then appeared in the dream and mentioned building blocks. Then the A-E column became a person whose slip was too long and was hanging below her dress, a border of lace being visible.

The second dream was simple though equally enigmatic. I dreamed that $6 + 1 - 5 = 3$. I really believed that this was true and it was proven in the dream by the fact that next not two but three of my neighbors came by and each of them gave me a halter while I stood in my doorway wrapped in a blanket. The dream then claimed that it was obvious that 4, not 6, + 1 was supposed to equal 5 and that the difference between 5 and 3 was 2, but by this time 5 had become f and four also f, six s, one o, and, in the ultimate mix-up, both 2 and 3 were t. Therefore:

$$
\begin{aligned}
\text{if } & s + o - f = t \\
\text{and } & f + o = f \\
\text{then } & f - t = t \\
\text{and } & t = t!
\end{aligned}
$$

Freewriting

Writing for three to five minutes anything that comes to mind, usually at the outset of a class period. This writing is intended to be private and can even be filed away or discarded, though not for lack of value.

Related to stream of consciousness, automatic writing, free association, and dayboooks, the process of freewriting has many beauties and advantages in the science classroom: it focuses attention and creates a moment of quiet in a sometimes noisy day; it loosens up the students' and teachers' writing abilities; it's good practice; it introduces the idea of writing for its own sake; it's fun, full of surprises; it doesn't have to be "correct" or "corrected"; it relates to mental processes important to scientists, such as dreams and daydreaming; it can help to solve problems; it can be useful to student scientists later in unearthing unnoticed ideas; it connects the act of writing with our thoughts as a whole. Freewriting is more like running than walking, and as such seems especially inspiring both to intellectually and physically athletic types, who might feel exhilarated by putting pen to paper and racing forward non-stop with any and every thought, idea, image, or question. Freewriting is an excellent workout for clear thinking.

If students become interested in this form of experimentation, freewriting can be extended for a longer period of time. Freewriting practice can also be delineated as "directed freewriting," that is, writing on a particular subject, problem, or idea, the writing then to be shared. Summary writing, often done at the end of a class, can also be useful to invite speculation. Another freewriting experiment is to write only in the form of questions.

Freewriting, though an age-old idea, is now one of the most current, frequently used, and most important experiments relating to writing-across-the-curriculum and the practice of science writing in school. It is discussed further in two other parts of this book—the Essay Development Workshop section and the Questions and Answers section.

Here is an example by a student who had nothing and everything in and on her mind:

Nothing in My Mind

I have nothing in mind, nothing in my mind, nothing in my mind, nothing in my mind, nothing in my mind, nothing in my mind, nothing in my mind, nothing in my mind, nothing in my mind, nothing in my mind, there's a lot of noise in the halls, all you hear is noise and Mr. M. saying "Hey get in line, get in line" so that's why there's nothing in my mind, nothing in my mind, nothing in my mind. Ms. Mayer is talking, nothing in my mind, nothing in my mind, Ms. Mayer told me her name, nothing in my mind, there's a lot of squeaky noise in the class, nothing in my mind, nothing in my mind, Marjorie and Nelsie are talking, nothing in my mind, nothing in my mind. Marjorie stood up, nothing in my mind, Joanni asked somebody "Do you have a problem?" Nothing in my mind, nothing in my mind. Ms. Maggs said, "I suggest you don't!" Nothing in my mind, nothing in my mind, nothing in my mind...

—*Sandra Estevez*

How and why we work to solve a scientific problem

An exercise related to freewriting. The teacher can give examples of "how": by visualizing, by beginning to write, by explaining, by continuing to think, by creating new meanings even if they seem abstract or irrelevant, by formulating questions, by further study, and by collaborating with others.

Why do we want to know the answer? Why do we want to discover something? Why do we want others to see things in a new way? Is there an answer that is "right" or "correct"? Is there no answer? Could writing or seeing things in a different way provide new possibilities for questions or conclusions? Why do we study? Why do we choose to study science? Why do we study science in a school? What is our intention in being scientists?

The experiment can be conducted as a ten-minute freewriting period at the end of a class.

Imaginary histories of science

Imagining and putting into words a scientific discovery of the future (or of the past) and how it came about, with a biography and history of the ideas of the imaginary scientist(s). Other possibilities are to create an imaginary space program, or to describe imaginary worlds, including maps, architectural designs, plans for the running of schools, etc.

Interdisciplinary investigations

Any of many possible experiments in surprise, detail, metaphor, and matter, and the mixing up of things. For instance, in a class on astrophysics, bring in a short story that is unrelated to the topic. Then attempt to interweave the scientific ideas with the story to produce a second piece of writing. One way of doing this is to introduce the definitions and etymologies of all the scientific terms, following up every lead in the dictionary. Interdisciplinary detectives will ordinarily find connections, however abstruse. There's no need to analyze, just to write in a new light.

Bring visual art into the physics or chemistry classroom. Is realism more expressive of chemistry than of physics? How do modern architects enter the realm of mathematics? Or, architects enter the realm of modern mathematics? What is sculpture? Is it biological? What of the science of chaos? What is the art of chaos? Does art reflect newly discovered physical principles? Or anticipate them? Is everything "old" and reinterpreted? What about computer art? Make a slide show combining art and science depictions; for example, alternate abstract art and images of galaxies, paintings of buildings and molecular constructions, portraits and equations, etc. Compare electrical wiring diagrams (or the plans for water and sewer pipes) with paintings by Mondrian.

Make a movie or videotape in which silent visual images are interspersed with verbal descriptions over a blank screen. How to envision a galaxy? How to depict finite and infinite? How to make a "moving picture" of absolute value? How does the gregariousness of humans and animals relate to visual patterns? What do those patterns mean in terms of order and chaos?

Synesthesia, or the mixing up of sense impressions, can also lead to thinking about the question of what music is reflective of what scientific idea, or vice versa. What sounds of animals, people, or things remind us of the concepts of chemistry, math, or biology? How does dance resemble mathematics or physics? What about birds?

Here's the results of a very simple experiment in interdisciplinary association:

47

Composer/singer(s)	Animal	Part of the body
Beethoven	frog	lungs
Mahler	lizard	liver
Whitney Houston	rabbit	spleen
Genesis	dinosaur	femur
Mozart	newt	heart
10,000 Maniacs	anemone	clavicle
Bruce Springsteen	hippopotamus	toe

Place histories

Here's a way to have a lot of fun with natural history and to let yourself and your own life into it. Write about a place you know: a streetcorner, a pond, a shop, a phone booth, a riverbed, whatever. Bring to it everything you know or can learn about the place, from its most distant past to the most recent thing you can remember about it. If you haven't got the time or inclination to research all about the place's deep past, make it up yourself, but keep in mind the general sorts of changes that the earth has gone through in the last billion years or so.

You may want to divide it into lines, like a poem, or write one big paragraph of prose. It might also be fun to do this as a collaboration, with each student working on a different part of the place's history. Both the following examples are about a streetcorner in New York City.

First, I was fire. Then, I became grey rock. I bent down and I was tossed up. Africa bumped into me. I broke and I cracked, and I started to look like shattered glass. A river ran through me, and it cut me in two. It went away beneath the ground and I climbed up where clouds live. Rain stuck in my cracks and melted my rock to soil. Something green and red came up out of me and it swung back and forth in the wind. A big bug, big as a bicycle landed on it. My tall back relaxed and I sank down again, and the river came back to me. Black fish with whiskered snouts swam over me. The river moved to someplace else, and I found something huge and grey that liked to stand on me. Then came the first cold, and the second, and I had a deep white blanket, and I slept while it carved and sculpted me. I threw off the cover and skinny two-legged things came to me. They lit red fires; they cooked by throwing hot rocks into baskets. Then there were others, and they took black tar and poured it on me. And made big houses beside me, and dug trenches through me. Last night, a woman walked up and down on me asking for help. And now the telephone in the booth is ringing.

•

It is 450 million years ago.
The solitary corals are here.
They are upset.
The reason is that trilobites have come.
Trilobites eat solitary corals.
That's ok. Solitary corals are nothing but tiny mouths anyway.
Let them get eaten for a change.
It is 250 million years ago.
Surprise! There are a thousand ants!
It is 50 million years ago.
There are big seeds and little seeds.
The little seeds make great big palm trees.
The big seeds make tiny reddish grass.
Who can believe it?
It is 20 million years ago.
The land takes off its heavy coat of ice.
The coat that kept it cold instead of warm.
It is 20,000 years ago.
Women sing Dio He Ko, Dio He Ko,
which means Corn Beans Squash Corn Beans Squash.
Raccoons nibble on the ripe ears.
It is 200 years ago.
A person has sold land! A person has bought it!
What does this mean?
The Iroquois don't understand.
Does anybody understand?
It is 20 years ago.
Here lands a little glowing flake, smaller than a gnat,
from the atomic test explosion in Nevada.
It is 2 years ago.
The Korean children are kicking a ball.
There's a woman in 3 coats who needs money.
It is 2 hours ago.
The hermanos sold a bottle of that green drink
and Mr. Garcia is standing here tipping it back.
It is 2 minutes ago.
The wires hum in my street.
And now
a telephone is ringing.

The question of one's relationship to the universe, the question of one's relationship to the environment, and the creation of federal and personal budgets: a series of writing experiments

The question of one's relationship to the universe

A writing experiment to attempt to situate ourselves as earth's inhabitants in terms of size and place.

A typical beginning might be to write your name, size, address, and galaxy, like a heading. A short period of meditation or meditative free writing might follow, concentrating on where and how one exists as an individual and what one's existence looks like in relation to the perceivable whole. Illustrations can be useful. It's interesting to combine this exercise with the writing of brief, rapid autobiographies. One possibility is to begin the class with a ten-minute period of autobiographical writing with the exhortation that, as in freewriting, students and teachers can write anything about their own lives that comes to mind — it does not have to be chronological. A discussion can ensue, followed by an attempt to answer the question: what is my relationship to the universe?

My students have given these answers to this question: risk; it's a big world and I am only one person; being alive is being part of the universe; I am one of many alien life forms; there are uncivilized people making matters worse for our living; drugs; when I grow, the universe grows also; I don't bother anyone and no one bothers me; I enjoy every minute of my relationship; McDonald's has 39¢ hamburgers on Valentine's Day; everything revolves around everyone, not just me; I don't know my relationship to the universe; I am a dot; I am a student now; I would like to work for NASA; the universe is big compared to me; I want to be an educator; I want to make things better; I am a cell, the earth an organism; my thoughts on this are jumbled up; I believe in family love; I want to do something and not nothing after finishing school; I want to be a free person; I hate the way this place is run and the people who live here.

Here are some fuller examples:

> My name is Yvon Mars. I was born on June 16 of 1972, so that makes me fourteen. I live in Manhattan, New York on 310 East 113 Street and Second Avenue. I now go to a high school called Manhattan Center for Science and Mathematics, which is a specialized high school. My goals are to become an archaeologist or a teacher in academic comprehension, because I love science and I like teaching. My relationship with the universe is that when I grow, it grows also.
> — *Yvonne Mars*

For this paper, I will define the universe as my immediate vicinity. Therefore, my universe consists of my home, school, other buildings, stores, and other things you would see in a common neighborhood.

My relationship to my universe is one of very great dimension. Everything that happens in it will affect me. I was born and raised in this universe and now I work and go to school in the same universe.

—*Anita Miles*

Hi! My name is Janet Arroyo. I'm sure it doesn't matter where I was born, I'm here and that's it. I am sixteen years old and by the end of June I'll be seventeen. My relationship to the universe is a difficult question to answer because I envision that relationship differently depending on my moods. At times I feel that atom size is still too large an object to compare the relativity between myself and the universe. Today I feel that the universe depends largely on how I wish to view it. I can say the four walls of my room are my universe. . . .

—*Janet Arroyo*

The question of one's relationship to the environment

Following the question of one's relationship to the universe can come the question of one's relationship to the environment. The word *environment* comes from the Middle English word *enviroun*, meaning around, or round about.

Writing around or "round about" our environment could follow the structure of ranging from the most immediate to the most distant. One's chair, desk, other living beings, the room, nature, the street, house or building, the store, the neighborhood, the town or city, the state, the country, the earth, the galaxy, the universe, and beyond that.

In doing this experiment, some students overlook the most immediate environmental questions, which is OK. For example, a concern with the plight of the whales might take precedence over what is happening ecologically in our own neigborhood. It's vital, however, to permit ecological questions to retain their entire range.

A good writing method is to suggest a series of three-minute writings on ecological questions, beginning with the self, the immediate environment, etc. and ending with the universe. Students seem to enjoy the rapid expansion of the subject. In effect, such a piece of writing enlivens the rather dull dictionary definition of ecology ("the branch of biology that deals with the relationship between living organisms and their environment").

The creation of federal and personal budgets

This experiment is an offshoot of studying ecology. Many of my students are uninterested in ecology as it is presented in textbooks, tests, and dittos, but everyone is interested in questions of money, budgets, and the personal power exercised in their management.

It's interesting to begin this class with some freewriting about money. Then, have students create personal, state, or federal budgets on the basis of limitless or limited funds. I have found limitless more fascinating to all. Along the way, it's fun to tell students the origin of the word *buck*: the dollar became a buck because it was the price of one deerskin in frontier days. You'll find that budgets are at the heart of ecological questions.

Here's the *World Almanac*'s list of personal expenses:

> Food, clothing, jewelry, rent, utilities, medical care, tobacco, alcohol, transportation, recreation, education, religious activities, and foreign travel.

and federal outlays:

> Defense, interest on the public debt, atomic energy, foreign aid, space and technology, science and research, space flight, energy, water resources, conservation, recreation, pollution control, farm income stabilization, post office, mortgage credits, transportation (ground, air, water, other), community development, disaster relief, education, social services, health services and research, retirement and disability insurance, unemployment, welfare, veterans benefits, law enforcement, salaries of legislators and government executives, and rent and royalties on the outer continental shelf.

Both personal and government budgets can be written either as lists (to be added and balanced) or in the form of essays.

Sending messages to other planets

What do we want to say to the other living beings of the universe? One student wrote:

> Greetings, aliens. I am an Earthling and I come from Earth. I came here to explain my existence and how I and my race came to be.
>
> It all started millions of years ago. A devastating explosion called the "Big Bang" brought all the planets together. By chance, up until now, our planet was the only one which housed living organisms which advanced throughout the ages. My race was once said to be an inferior race of sub-humans called apes. This, of course, was all a theory. Maybe we can collaborate all our knowledge and create a superior race which can rule throughout the universe.
>
> Together, Alien and Human can reach the end of the cosmos.
> —*Rafael Fernandez*

Another possible experiment is to create an abstract way of communicating with all possible forms of living beings. See Peter Sears' *Secret Writing* for possible codes to be used in sending messages into space, including the first message sent into space (in 1974 from the Arecibo radio telescope in Puerto Rico), examples of pictorial and numerical messages, ciphers and codes, messages based on binary numbers, the atomic numbers for hydrogen, carbon, nitrogen, oxygen, and phosphorus, the formulas for sugars and bases in nucleotides of DNA, and the number of nucleotides in DNA, and the engraved plaques or "cosmic greeting cards" attached to Pioneer 10 and 11, the first spacecraft to leave the solar system. These plaques contain depictions of male and female earthlings, information about the location of the sun and of the earth in relation to other planets, an image of the path of the Pioneer spacecraft, and information about life on earth.

Sensory perception

A traditional warm-up writing exercise in memory and objective description. Invite everyone to write two pieces: first, record all sensory memories (smell, sound, taste, touch, sight) from breakfast or lunch; second, record perceptions of the present moment, based on observations of the classroom, or what is out the window, or who is near, or what one's hands look like.

Perform each experiment separately, pausing in between to delineate the difference between working with the inside (memory) and the outside (sensory perceptions); or perform them alternately.

Thirteen ways

An experiment inspired by Wallace Stevens' poem "Thirteen Ways of Looking at a Blackbird." In any writing form, write thirteen different ways of perceiving or approaching a topic that is relevant to the science class. The more specific — or even tiny — the topic the better. You might present Stevens' poem and discuss the idea of divagation, or wandering about, straying from the subject.

Here's an example, written by a ninth-grade student, of five ways to look at the letter *A*. Franzy's use of the simile is only one of a myriad of possibilities.

1—An "A" is like an intergalactic space vessel, if you have a three-dimensional view of it.

2—An "A" is like an underwater mini-sub which explores the ocean floor, if you have an overhead view of it.

3—An "A" is like a tip of an arrow if you put a stick on the middle part.

4—An "A" is like an airplane steering wheel if you connect the back of it to a stick which in turn is connected to the airplane's main computer terminal.

5—An "A" is like a space age aerodynamic futuristic car which hugs the road as it goes at speeds up to 700 mph.

—*Franzy Belfort*

Word equations

Have students write equations that use words instead of numbers. For example, instead of

$$2 + 3 = 5$$

they could write something like

grandmother + dream = interstellar space.

Or more complex relationships:

The square root of desire is equal to the body minus time.

A variation is to have students write parodies of algebraic word problems:

If Dick is rowing upstream toward Chicago at the rate of 1 mph, and the stream is flowing at the rate of 3 mph, and the wind is blowing from the NE at 5.2 mph, and Dick pauses to rest his poor arms every 1.5 minutes, how many sandwiches will Dick have to eat to maintain his weight, assuming the work he performs causes him to lose 1 lb. for every mile he rows, and each sandwich adds a quarter of a pound to his weight? And why is he rowing toward Chicago?

Another variation is to take chemical formulas and "translate" them. For example:

H_2O = Hurry to Oregon

$NaCl$ = Navaho Clorox

H_2SO_4 = Happily, two screaming oblongs forgot.

Students should then use these formulas as the basis for longer pieces; e.g., the screaming oblongs could be characters in a science fiction story.

Rewriting textbooks

For many high school students, this is the first experiment in structuralism, which is a word for the modern intellectual method of reweaving the whole picture by one's own self.

From the class textbook, have each student choose a section or chapter that he or she feels is very interesting or quite dull, and then rewrite it. Encourage new forms and approaches: the selection can be made more concise; its information can be presented in the context of one's personal experience; new charts, diagrams, and illustrations can be created; the language can be made less formal. Allow students to take any approach they want, including parody and satire.

Unlike many types of books, textbooks are relatively anonymous; they "belong" to all of us. Any student who knows his or her subject well can write a textbook, and in fact students should write them, to impart their knowledge in a more direct and personal way, and to spread the idea that the sharing of knowledge in writing need not be restricted to the so-called adult point of view.

In some good schools, the older students have partners in the lower grades — seventh graders and second graders, for example — but in any school it is good for the older children to write books for the younger ones.

Other possibilities for rewriting textbooks are to write additional chapters on neglected material or to provide new introductions or afterwords. Tests should also be created by the students. Figuring out which questions to ask, whether to include essay questions, multiple choice, or other formats — all these decisions bring the students to the subject material from a fresh angle, allowing them to understand it in a different way.

Science projects for the future

The discussing, imagining, outlining, and budgeting of conceivable and inconceivable scientific projects, solutions, and experiments, which might include: discovering if the universe is finite, finding a cure for AIDS, making math of the patterns of sand dunes, making the confluence of the flower with the bee disorderly, finding the ideal shape for a glass or a table, etc.

The question of human reproduction

Remembering that science and sex derive from the same root word, the Latin verb *secare*, to cut or divide, which accounts for Webster's peculiar definition of sex: the divisions into which people, animals, or plants are divided, with reference to their reproductive faculties. After which we are invited to see *saw* (related also to cutting and dividing).

I've had the temerity to give this assignment only twice, because, as one student wrote, "Sex is a very 'touchy' subject." I handed out a xeroxed anthology of articles relating to current scientific and ethical questions of human reproduction: "safe sex," government interference in private life, surrogate motherhood, experimentation on human embryos, abortion, test-tube fertilization, sex education, STD testing, pregnancy detection, birth control, etc. We had a lively discussion and then wrote poems and short essays, many of which were well constructed and carefully thought out. Here's an example in the acrostic form:

Sex and Politics
Symbol of human nature,
Evolve around my heart and soul
X-plain the mysteries of the Deity's work.

And
Now
Destroy this planet with this.

People make the world go round, sometimes it loses control
Oops, another error. I know I'm still human too.
Live this life very merry
If not so, it soon will tarry.
To all those who think they live in the clouds,
Instead, like ostriches, they hide underground.
Could we be wrong, or maybe right
Someday we will all know.
    —*Hank Bueno*

Permutations in writing

What we hear is often random, and in terms of astrophysics the universe is quite unpredictable, even chaotic.

Have everyone reorganize or reorder science or mathematics material at random, as an attempt at the total transformation or complete (chance) rearrangement of any given, based on the idea of interchanging. This is an experiment in discovering whether permutations create new ideas. If 1, 2, and 3 taken two at a time can be 12, 21, 13, 23, 31, and 32 (six combinations), then words, lines, phrases, and sentences put together in random combinations might result in something interesting too.

In music, John Cage and other composers have made use of random methods that include computers; the throwing of dice; the methods of the ancient Chinese text, the *I Ching*; incorporating everyday sounds in composition; and chance techniques to determine, on stage, in what order and way a work is to be performed. Painters such as Jackson Pollock also used randomness by splattering canvases with accidental dots, splashes, and patterns of color reminiscent of the visual data of electronics and astronomy. The poet Jackson MacLow incorporates chance into both the reading and writing of his poetry, through the random introduction of words and phrases from outside sources, as well as different reading methods, rendering the work different every time it is seen or heard. Of course, at the moment of the creation of a work through aleatory methods, its new meaning may not be apparent.

Writing techniques that can be used in the classroom include: making the last sentence be the first, the penultimate the second, etc.; numbering a poem or essay's sections from 1 to 6 (or 12) and throwing the die (or dice) to determine a new order; combining the first word and the last words of each typed or written line to discover what new combinations of thoughts eventuate; cutting pages in quarters and replacing one quadrant with another; finding clues from numbers that appear at random, such as the temperature; taking all the words and phrases that "stand out" in a given piece of writing and making a list of them to discover why they seem important; repeating things that "stand out"; combining two or more people's writings on the same subject by interspersing paragraphs, phrases, or ideas; reading a text backwards; and inventing new chance methods of one's own.

When dealing with the operations of chance, simple methods create results as magical as complex ones. And what about 11, 22, and 33? Repetition can be explored mathematically and verbally.

Here is an excerpt from a verbal example of repetition called "If I Told Him, a Completed Portrait of Picasso," by the early twentieth-century experimental writer, Gertrude Stein:

> If I told him would he like it. Would he like it if I told him.
>> Would he like it would Napoleon would
> Napoleon would would he like it.
>> If Napoleon if I told him if I
> told him if Napoleon. Would he like it if

I told him if I told him if Napoleon.
Would he like it if Napoleon if Napoleon
if I told him. If I told him if Napoleon
if Napoleon if I told him. If I told
him would he like it would he like
it if I told him . . .

Presently.
Exactly do they do.
First exactly.
Exactly do they do too.
First exactly.
And first exactly.
Exactly do they do.
And first exactly and first exactly.
And do they do.
At first exactly and first exactly
      and do they do.
The first exactly.
And do they do.
The first exactly.
At first exactly.
First as exactly.
At first as exactly.
Presently.
As presently.
As as presently.

Repetition jogs us into thinking of words and ideas in new ways.

A history of one's own ideas

When Albert Einstein was asked to write his autobiography, he wrote little about his personal life and mainly about the history of the development of the ideas that led to the general and special theories of relativity and to his other conceptions in physics.

Students and teachers can set aside a day or a week or a year to attempt to write a history of the development of the scientific ideas that have influenced them most, how and why, and what might happen in the future. Such histories or autobiographies of ideas most often will be written in the form of discursive prose, which may be interspersed with diagrams, equations, illustrations, and other visual data. Occasionally a history of

58

an individual's ideas has been written in poetic form, for example Wordsworth's *The Prelude* and R. Buckminster Fuller's *How Little I Know*, two long poems. The work need not be long, however.

When Einstein wrote his history, he spoke abstractly about the concepts of thinking and wonder, and about the feelings of awe experienced in childhood. In writing such a history, it's vital to recognize that the way we think, learn to think, reach conclusions, and create questions are as much the stuff for analysis as what we know (and what we do not know).

A good way to begin this project is to discuss childhood memories, especially those that relate to wonder and awe.

Mutual aid

This involves writing and thinking about the evolutionary concept of the sociability, mutual protection, and shared struggle for existence of ants, bees, bird, and humans in primitive tribes and in cities.

It's fun to begin this experiment by bringing an ant colony to class and then asking "Why do we live in cities?" The writing exercise can be based on direct observations or on memories of mutual aid among human beings. Here is an example:

> Organisms help each other so that they may survive and evolve faster. Yesterday I had to babysit in Manhattan. I had to take care of two kids, one is four years old, the other twice her age. I had a problem getting them to go to sleep because they both refused to go to bed on time. Both of them fell asleep at about ten thirty and I went to bed an hour after that.

This is a good experiment to use after discussing Darwin, since, in high school today, the concept of mutual aid creates as spirited a set of disagreements about human nature as the theory of evolution creates about the relationship between religious and scientific belief.

Peripatetic scientists

The word "peripatetic" comes from the Greek *peripatein*, to walk around, and from the Indo-European base *pent*, "to step" or "to go" (related to the words "find" and "bridge"). It refers to the followers of Aristotle, called peripatetics, who walked about in the Lyceum with him while he was teaching.

Have students take a walk together, discussing scientific matters, observing, and making notes about everything. The scope of discussion can be limited to the sizes of things, the colors of things, types of trees, questions concerning the construction of cities, kinds of materials and stone observed, the weather, and so on. Written records can be gathered and compared in the classroom. The ambience of this experiment can be either intently serious or lighthearted and hilarious. It seems to work either way. Stress the importance of detailed observation.

Writing in poetry form

The uses of songs, sonnets, free verse, acrostics, and other poetic forms for the elucidation and exploration of scientific material. As William Wordsworth said: "Poetry is...the impassioned expression which is on the countenance of all science."

Set aside a class that can be entirely devoted to writing (and reading) the work that is written. Explain the poetic forms you think would be appropriate for a current class idea. Songs require no explanation, except to say that they need not adhere to any strict metrical or rhythmic schemes, as is true of sonnets, which can simply be described as fourteen-line poems that have a tendency to reach a conclusion. Acrostics seem adaptable to all subjects. (See science acrostics in this Writing Experiments section.)

In one ecology class, we obtained a set of cards with pictures of life forms on one side and verbal information on the other. We each chose one or two and created acrostics based on the names of the animals, making use of the scientific information given. For example:

Sea Lion
Sea lion swims upon the beach
Eating fish and swimming free
Attacking each other. The males
Lingering behind a rock
Inventing new ways to defeat the enemy
Opening and closing their jaws
Not caring about their fellow males
    —*Anonymous*

Ghost Frog
Green
Heleophryne rosei
Ordinary location in the south of Cape Province
Skin so white its organs can be seen
Transparent
Females that are 20 mm larger than the males
Rivers that carry the females' eggs
Ovary that holds 30 eggs
Great resistance to strong currents
    —*Anita Myles*

60

A longer form, the sestina, is ideal for people who love complexity and/or mathematics. The sestina is a thirty-nine-line poem consisting of six unrhymed stanzas of six lines each in which the words at the ends of the first stanza's lines recur in a pattern at the ends of all the other lines. The pattern is created by using the last word of the previous stanza first in the following stanza, the first second, the second last third, and so on. The sestina concludes with a tercet (three-line stanza) that also uses all six end-words, usually two to a line. In the diagram of the sestina form below, the letters A–F stand for the six end-words of the poem:

Stanza 1: A
          B
          C
          D
          E
          F
Stanza 2: F
          A
          E
          B
          D
          C
Stanza 3: C
          F
          D
          A
          B
          E
Stanza 4: E
          C
          B
          F
          A
          D
Stanza 5: D
          E
          A
          C
          F
          B
Stanza 6: B
          D
          F
          E
          C
          A

```
Tercet:    AB
           CD
           EF
```

You can begin to write a sestina by choosing the six end-words first, or by writing six lines and seeing what the end-words turn out to be. Once you know what they are, you might want to write down the pattern they'll create for the rest of the poem, so you won't have to be distracted by puzzling over which word is supposed to end the next line.

Here is an example in which the end-words are *space, robot(s), technology, wires, computers, microchips*. This poet has taken some liberties with the form, which is fine; he's exercising his "poetic license."

Sestina
Up there, there's space
Something controls robots
I think this does it—technology
Robots have wires
Missiles are controlled by computers
And computers are controlled by microchips

I love microchips
They got a lot to do in space
They fit in a computer
They control any robot
Microchips are connected to the wires
All this has got to do with technology

Humans made technology
Technology created microchips
Microchips need wires
With all this, humans discovered space
But something took over, robots
But we got defense, computers

This makes humans lazy—computers
But this made us work again—technology
But this made us lazy again—robots
And this got even worse—microchips
But microchips took us to space
To fix some old dumb wires

Robots wouldn't operate without wires
Wires made this useful—computers
Wires got us into going to space
Wires made this possible—technology
But sometimes wires need help from the microchips
So wires now made the robots to live

I love robots
I like wires
I like microchips
I love computers
I like technology
But I don't like this place because there is no place
    to play with robots, no place to have a computer,
    no place to put microchips and place is space!

Oh space and robots
Technology and wires
Computers and microchips!
   —*Rene Munoz*

Experiments with dailiness

Have students write on subjects such as the science of cooking (both at home and in the laboratory); how guns work; what ballpoints are; how jets go; what languages are; how color TV works; how a flash bulb works; why a ship floats; what glass is; how gas and water meters work; how an automatic transmission works; what a quartz clock is; what the difference is between a compression refrigerator and an absorption refrigerator; what Plexiglas, enamel, rubber, and porcelain are; why and how toilets and door locks work and don't work; and what a differential gear is. Good references for such subjects are *The Way Things Work* (no author credited) and *Extraordinary Origins of Everyday Things* by Charles Panati.

    Everyone can present subjects in the form of questions, answers, or suggestions, then research and write about them in the form of essays or poems that are both ruminative and factual. (See the William Carlos Williams poem in the Samples section.)

Experiments with sound

Echo was a nymph in Greek mythology. Because of her unrequited love for Narcissus, who had fallen in love with his own image, Echo pined away until only her voice remained. Another version of the myth says that Hera, Zeus's wife, deprived Echo of her power of speech unless she was spoken to first, and then she was compelled to reply with the same last word that had been spoken. Hera did this because Echo was trying to distract her, by incessant talking, from Zeus' dalliances with the other nymphs.

Echoes, sonar, and radar are forms of reflection and repetition, ranging from bats to whales to poetry to mountain yodelling. *Echo ranging* is the generic term for obtaining data about distant objects by the transmission of energy and the reception of its reflection.

SONAR means SO(und) N(avigation) A(nd) R(anging). RADAR is an acronym of RA(dio) D(etecting) A(nd) R(anging).

Excellent and mysterious poems have been made by humans (and maybe whales and dolphins!) using echoes. Here are some examples by humans.

If I address the Echo yonder,
What will its answer be I wonder?
   (Echo) I wonder.

O, wondrous Echo, tell me, bless,
Am I for marriage or celibacy?
   Silly Bessy.

If then to win a maid I try,
Shall I find her a property?
   A proper tie.

If neither being grave nor funny
Will win the maid to matrimony?
   Try money.

If I should try to gain her heart,
Shall I go plain, or rather smart?
   Smart.

She mayn't love dress, and I, again, then
May come too smart, and she'll complain then?
   Come plain then.
     —*Excerpt from anonymous 17th-century poem*

From "An Echo"

Ah, will no soul give care unto my mone? one.
Who answers thus so kindly when I crie? I.
What fostered thee that pities my despair? aire.
Thou babbling guest, what know'st thou of my fall? all.
What did I when I first my faire disclos'd. los'd.
Where was my reason, that it would not doubt? out.
What can'st thou tell me of my ladies will? ill.
  —*Lord Stirling (1604)*

Heaven
O who will show me those delights on high?
    Echo. I.
Thou Echo, thou art mortal, all men know.
    Echo. No.
Wert thou not born among the trees and leaves?
    Echo. Leaves.
And are there any leaves, that still abide?
    Echo. Bide.
What leaves are they? impart the matter wholly.
    Echo. Holy.
Are holy leaves the Echo then of blisse?
    Echo. Yes.
Then tell me, what is that supreme delight?
    Echo. Light.
Light to the minde: what shall the will enjoy?
    Echo. Joy.
But are there cares and businesse with the pleasure?
    Echo. Leisure.
Light, joy, and leisure; but shall they persever?
    Echo. Ever.
 —George Herbert (1593-1633)

Here's a contemporary and deliriously whimsical example:

From "The Echo"
Violets! No flower can compare
        Pair
With your frail beauty
      Yooty
As I bend down to sniff you
        If you
I must declare
    Air
That such frail power
      Hour
Resides in your frail beauty
       Ooty
That like a frail bunny
      Unny
I must beware
    Where
Lest by some chance
     Ants

```
I too beguiled
              I'ld
Should rest, should stay
              A
Here by your fragrant bosom
                        Uzzim.
    —Kenneth Koch
```

Some classroom experiments could include:

• Write an echo poem from the point of view of any combination of the following: an average person, a nymph deprived of all but echoic speech, an alien, a scientist, a whale, etc.

• Write a brief essay on what is sound and how does it return to us? Also, what is noise?

• Write about the noises you hear right now.

• Listen to them and imitate them vocally and on paper. (Echo them.)

• In an essay, attempt to communicate with other species.

• Visit an aquarium (or buy a record) and listen to eels, whales, and dolphins. Translate their speech. What time is it for them? Do they know you? Who is visiting whom? Can whales' language be heard by us more clearly than ours by them?

Private journals

The keeping of a record of our classes in the form of a notebook, journal, or daily ledger.

Time can be set aside at the ends of classes to write in such journals, which are neither handed in nor graded. Doing the work for its own sake gives it a deeper value. The teacher keeps a journal too. Students find the keeping of a journal an unexpected delight—a sort of vacation—but one that also provides both students and teachers with a kind of review more lively and meaningful than note-taking. In the course of journal writing, more questions are posed, ideas are related to one's own experience, and apparently irrelevant material enters in, often finding meaning in the end. In keeping journals, the ancient spirit of inquiry, combined with the benefits of having no audience, gives teachers and students access to an informal method of learning free of rigidity and the misguided quest for instant knowledge. Journals enable people to become better problem solvers.

Journals can also incorporate charts, diagrams, and illustrations.

Science acrostics

The acrostic is a poem in which the first letter of each line, read downwards, form a word, phrase, or sentence. The subject of each poem can relate to its "spine-word," or not. In the first three acrostics below, the vertical word is "science":

Science seems to
Come into every
Introduction to
Every last
Notion at the Manhattan
Center of
Entropy

Science places
Calm clams
In
Every Ocean
Nothing is
Critical but
$E = MC^2$

Science is
Corrupt yet maybe
It is not.
Everything
New
Contains
Everything
   —Anonymous

Here are some examples for *atoms, math, philosophy, absolute value,* and *factorization.*

All
Tiny
Objects
Make
Sense
   —Rodney Pink

Maybe the most
Astonishing subject,
Terrorizes and is
Hard.
   —Yesenia Ramos

People
Have
Incredible
Love
Of
Space
Other
People
Have
Yaks
    —*Eurik Perez*

Absolutely un
Believable the way
Some people understand the value
Of math just
Like adding of integers it's really
Unbelievable
The way you know what you're saying
Even if you don't

Valuable math
And
Like addition of integers
Understanding a person's language
Even in English
    —*James Rivera*

Fast
And
Complicated
The
Others say
Really not hard but not easy either
I think the way they think
Zooming around around
And
Thinking
It
Over and over again
Nagging about the things they know
    —*James Rivera*

Acrostics can also have the vertical word at the end of the line or in the middle. This is an especially useful form for developing verbal inventiveness. In science classes, it is marvelous to see metaphor lead to new ideas and speculation about scientific subjects through the use of the acrostic.

The use of etymologies

In class, make frequent use of the origins of science words such as *atom*, *decimal*, and *epiphysis*. Students and teachers should rapidly get out their dictionaries to find the unique clues to understanding that etymologies often provide.

Here are some examples:

- muscle: from the Latin *musculus*, meaning "little mouse."
- science: from the Latin *scire*, "to know, to cut, to divide, to separate" (related to "skill" from the Swedish *skal*, meaning "reason," and the Icelandic *skilja*, meaning "it differs"; also related to "sex" from the Latin *secare*, "to cut or separate"), and from the Indo-European root *skei*, related to "scissors."
- mathematics: related to the Greek *manthanein* ("to learn, to be alert"), the Indo-European *meudh* ("to pay attention"), the Persian *mazda* ("memory"), and the Geman *munter* ("cheerful").

No wonder mathematicians have such good memories, while being cheerful, alert, and attentive. No wonder scientists are always dividing things with their reasoning scissors!

- technology: from the Greek *technologia*, "a systematic treatment," which in turn derives from *techne*, "art or artifice" (from Indo-European *tekton*, "carpenter") and *logos*, meaning "word or science."
- An "engineer" (a producer) means much the same as "poet" (a maker), but etymologically speaking, the engineer has the better of it, being derived from the Latin *ingenium*, related to "genius."

Basic etymologies can be found in most editions of *Webster's Collegiate Dictionary*, in Eric Partridge's *Origins*, in *Skeat's Etymological Dictionary*, and many other etymological dictionaries.

Rube Goldberg

Show students a Rube Goldberg drawing, one of his funny fantastic home-made devices that goes through numerous complex steps just to do something simple, such as turn on a light. Then have the students draw their own devices and describe them with a long caption.

You might also show them the amusing *Absolutely Mad Inventions* by A.E. Brown and H.A. Jeffcott, Jr.

An attitude of silence to the stars

This poem by Walt Whitman expresses the simultaneous doubt and wonder of the student (and teacher) of any science:

When I Heard the Learn'd Astronomer
When I heard the learn'd astronomer,
When the proofs, the figures, were ranged in columns before me,
When I was shown the charts and diagrams, to add, divide, and measure them,
When I sitting heard the astronomer where he lectured with much applause in the lecture-room,
How soon unaccountable I became tired and sick,
Till rising and gliding out I wander'd off by myself,
In the mystical moist night-air, and from time to time,
Look'd up in perfect silence at the stars.

Study the poem, thinking of the beauty of natural events and of the gorgeousness of scientific ideas, then imitate the poem by writing a piece that begins, as the poem does, with four "when's," followed by "how..." and "till...."

When preparing to write, think of the idea of silence in combination with studying and teaching and writing. Think of why we choose to express some things and not others. How much do we know? What have we seen?

Should we then not investigate, put things into words or formulae? Is science dull and unrelated to the stars, the clouds, the weather, the fruits and vegetables we eat, to sex and love, to evolution, to the complex matters of death and wealth, to the daily matters of where our water and electricity come from?

The piece could end with a sentence about silence — how it can be "perfect."

Poems using selected scientific words

Write poems that use certain words, for instance, a poem with *battery, direct current, circuit breaker, atom,* and *magnetic field.* It's good to use free verse for this experiment, and to have at least several lists available. Example:

I flow along the direct current
to your battery, but your eyes
are circuit breakers that cut the flow
of my atoms into your magnetic field.

Writing about atoms

Prepare an information sheet about atoms and molecules, including definitions, illustrations, and etymologies of atom, bond, compound, molecule, chemical bond, etc.

Using the information given as a taking-off point, write in fiction, poetry, or essay form. Here are some examples from a ninth-grade class:

> God, the father of Atom and Eve, also had more children, hydrogen and oxygen. You know the "real story" about the apple in the forest, but some believe another version.
>
> One day Atom and Eve were in a science lab. They decided to be experimental, so they took an oxygen atom and two hydrogen atoms and combined them to make a water molecule. After their experiment was made, the lab was filled with small children, grandmothers and grandfathers, mothers, fathers and any other family generations you could think of. Their conclusion was "this is how people get here."
> —*Keshia Windham*

> One day there was a kid named Oxygen. Oxygen was outside playing when the Hydrogen twins approached. They asked him for some money since they were dying for something to cool them off. Little Oxygen was scared. He had no money to think of. Just then the Hydrogen twins beat little Oxygen up, and in doing that, something that we now call water was flying away from them. It got so bad that all of a sudden the kids fell into a big dug-in dirt ditch and were never seen again. When you drink water, think about how these kids sacrificed their lives to give you our first beverage.
> —*Andre Sanchez*

Complexity of thought

Here's an attempt to create writing that reflects — yet does not speak "about"—the complex nature of thought. Invite everyone to write about the most complex scientific or personal topic they know of, one they feel cannot be understood. Give a variety of possible forms in which the writing could take place: freewriting; discursive writing; poetic forms; a visual image, as if one were drawing a dream; a design of words on a page singly or in phrases or sentences that seem best to reflect thought. Encourage both the visual and verbal creation of transitions between aspects of a thing or things, an idea or ideas.

Subjects for this experiment have included death, disorder, advanced algebra, nucleic acid, cute guys, philosophy, calculus, the news, and white dwarfs. Here are two examples:

The square root of 2 to the decimal comma by E = mc² to the supposed interest of man to space if a dehydration to the skin and to mind biology and biochemistry mix/do not mix with each other has me to the waterplow to a toilet bowl to a field of gumdrops up and down all around jump hop skip don't matter where you land fall up throw down dog meow and a cat barks he can they do because of me to he of me you see. The dreams of an unknown turtle to me to be again you see if we dream of monsters, creatures and things that'll go bump in the night a mouse dreams about a cat a cat dreams about a dog what does a dog dream about it's come to Christmas time for humbug and time for taking do you think I'm going to jump low, walk low, that's the way to do it. Pretend in a dream just jump into someone's hair and be in a forest.
        —*Eurik Perez*

        Dis ain't da way I should rite
        I should bee more intelligent
Maybe I Should learn more to make da world better
        Or should I stay da same and cauze more
    DisoRder
        I Did not mean to hurt u, i didn't know
        Excuse me if i cry.
        Remember mee cauze i'm not
                important but two me I'm
                        quite brite.
    —*Hank Bueno*

Positing wild theories

In objective science writing we cannot say something that patently isn't true, or present as truth something that cannot be verified or proven. In a literary essay or freewriting exercise, we can say "Human beings are filled with sadness" or "Snow comes out of the ground and flies upward" because these are acceptable as individual opinion or literary device. Nonetheless, it's challenging to say, in a scientific essay, something such as "An atom is the basic component not of matter but of memory. I will now support this hypothesis," etc. It is an interesting experiment to posit wild, instinctive theories and attempt to prove them in writing.

72

Found poems

From a science book or magazine, or from the newspaper, have students pick a sentence or paragraph they like and break it into lines, so it resembles a poem. Here's an example from poet John Giorno:

Astronaut Jim Lovell
flying in Gemini 7
high over Hawaii,
today spotted
a tiny pinpoint
of greenish-blue brilliance
far below.
He successfully "locked on"
for 40 seconds
and sent
the world's first communication
down a laser beam
to earth.

"I've got it," Lovell cried.

Giorno's selection and arrangement of lines conveys the intense excitement of that moment in a way that flat prose doesn't, even though he uses the same words as the newspaper.

There is an instructive scientific lesson here: a specimen, no matter how mundane or small, removed from its context and examined closely, becomes much more interesting. When it is returned to its context, the context becomes more interesting, too.

Epistemology

Epistemology comes from the Greek word *episteme*, meaning "knowledge" and *logos*, meaning "word or science." Though it seems formidable to experiment with the origins, nature, method, and limits of knowledge, this becomes the simplest of writing exercises, and works equally well for people of all ages.

Invite students to write a series of ten questions on any or all subjects, and then to write a second series of ten questions about the subject of the class. Then, in a third exercise, have them write about how to find the answers. For instance, what book or library would contain the information I need? If I don't know, whom could I ask? Where is the nearest bird sanctuary? What means of transportation do I use to get there? Can I call the public library or the natural history museum to find the answers to questions? Can I call a professor at a college or university? Can I call someone who works in a private business? Can I call or write the author of an article or book to get answers? Or even, how do I obtain the money to buy an expensive book I need? Finding answers often leads to further questions.

Though the questions need not be answered, a further experiment could be to exchange questions and attempt to answer each other's, then to return them to the questioner for comment.

If there's time, a good summary writing exercise could involve the idea of the questions themselves: How do we know what we know? Can we know everything? Do we know everything just by the fact of existing? Is it important to know the names of all things—rocks, minerals, fish, birds, mammals, stars, flowers, elements, trees, etc.?

A brief example comes from a ninth-grade student:

> I know everything I know
> Such as
> Education
> My religion
> The danger around us.
> —*Herman Zarate*

Porcus latinae

First, take the scientific (Latin) names of some beautiful/ugly creatures. Translate them and discuss how they were given their names. Then create your own scientific names for them, by concatenating English words that describe them. First, take more general characteristics, suitable to the Genus; then take more particular characteristics, proper to the species.
So, for example, we have the
Potbellystinkytailblackwhitestiffhair-us
spottedpinkeyedtwelvenippledlikestoclimbpinetrees-ae
or Western spotted skunk.
It might be even more fun to do this backwards, making up a huge "scientific" name in our heads, then letting the class try to draw it. The students could also make up scientific names for themselves in this fashion.

Writing about everything *but* the subject

In this experiment the teacher must "trick" the students by beginning a regular class on a subject or issue. After twenty minutes or so, including discussion, both students and teacher attempt to write for fifteen minutes about anything that *does not* relate to the subject. Set aside some time to read the works aloud, not only to have fun but also to create interest in the absent subject. This is a difficult experiment. Accolades can be given to the people who are able to avoid the initial topic completely.

Writing for a non-scientific audience

Write a science lesson, description, or article that is intended to be read by someone who knows nothing about the subject. Try to explain complex concepts simply and clearly. These pieces can be addressed, like letters, to one particular person, or written in dialogue, in the form of a phone call.

In the beginning of Herbert Kohl's book *Mathematical Puzzlements*, there is a quote from David Hilbert: "An old French mathematician said, 'A mathematical theory is not to be considered complete until you have made it so clear that you can explain it to the first person you meet on the street.'"

One good way to learn about any subject is first to read a children's book about it. Most people—especially adolescents—are embarrassed to do this.

Questions

Have students write nothing but questions, or let students pose and answer each other's questions without worrying about whether their answers are right or wrong. It's also fun for the students to write answers and have other students write the questions for those answers.

Anonymous questions permit students to forget pretense and ask the questions a younger person might ask.

In writing exercises with people of all ages, I've had great success with questions because questions do not seem to most people to be like writing: they are right there in the mind, instinctive.

Word ratios

Have students use words instead of numbers in geometric ratios. For instance, instead of

2:5::4:10

it could read

apple:fruit::Buick:car.

The ratio could then be expanded into

My new applecar is a Fruitbuick.

This is not only a good exercise in logical thinking, it's also a way to create startling images that could serve as the start of a poem or story.

Bee lines

Have students read Otto von Frisch or other material about how bees communicate with each other by dancing. Then have students think and write about a moment when they themselves communicated by dancing. Have them close their eyes and "see" that moment again, so they can give a specific, detailed, graphic description. (See Bernell Hollis's piece in the "Objective Observation" discussion in the Essay Development Workshop section.)

Uses of newspaper and magazine articles

It's fun to make collages, anthologies, and even cut-ups of recent newspaper articles relating to the class subject and then simply give them to the students, for no designated use. It's good to outline the sentences or sections that interest you most. The *New York Times*' Tuesday "Science Times" and the Sunday "News of the Week in Review" sections have good articles, especially on physics and ecology.

Magazine articles can provide examples not only of scholarly but also popular writing that is often lively and un-textbookish.

Möbius Sentence

Take a strip of paper (adding machine tape is perfect) about thirty inches long. Turn one end of it over and tape or glue it to the other end. If you start drawing a continuous line along the length of the paper, you'll find that the line eventually runs the length of both sides of the paper and ends where it began. This is the famed Möbius Strip—a piece of paper that has no front or back, no beginning or end.

Have students make a Möbius Strip and write on it a continuous sentence—one that has no periods, semicolons, exclamation points, or question marks and whose "end" blends perfectly into its "beginning." The "beginning," of course, shouldn't be capitalized.

Questions and Answers

INTRODUCTION

This section is a forum in which issues and ideas are discussed as if the authors were answering questions at a conference of science and math teachers. The format is somewhat freewheeling and the focus shifts between issues of the day and interesting concepts and ideas. The discussion may be entered at any point. Readers are advised to seek out the questions they themselves might ask. As a general rule, the subjects explored here have not been addressed elsewhere in the book. If they have been, reference is made to where they are elaborated upon.

Will using writing in the classroom increase the workload for teachers?

Many teachers, particularly science and mathematics teachers, might be interested in using writing in their classes, but are already burdened with too much paperwork to think of adding to it with reams of student writing. This worry is unnecessary. For one thing, students will become more interested in their subjects through writing, which will generate curiosity. A process will get started that will make learning more efficient, and the burden of unpleasant work on the teacher will actually be reduced. For another, the teacher doesn't need to read all the writing. The students can share it with each other and even correct each other's grammar.

Can students be trusted to do a lot of work on their own?

Many students are accustomed to teachers caring only about what it looks like they are doing. Here we care about what they are thinking. They are already thinking a lot. When we recognize this, they will think more, and what they do will not only seem appropriate but be appropriate. As a consequence, most of them will do their work without our having to worry about it. Additionally, the collaborative teaching methods implied by our writing ideas make student participation an issue shared by the class as a whole.

What if students say their work is finished when it isn't?

Tell them what you think and ask them to write an explanation of why they ended the work when they did. Take it from there.

What if students write too much?

There may be several reasons for students writing too much. If they are in the first draft of an essay, they may be including ideas that they will discover they need to cut in subsequent drafts. Through the process of peer criticism, another student can help them decide what to cut. They may have written too much because they are not paying enough attention to what is truly interesting in their work. Usually this is because they are not themselves interested in what they are saying. At this point, it is not your responsibility to correct the writing, but to help solve the problem of the students' lack of interest. Writers who appreciate the power of writing place a great deal of emphasis on the pleasure of it. Whenever it is becoming too onerous a task for the teacher or the student, something is wrong. A problem has arisen that needs to be solved.

What if students have bad ideas, or wrong ideas?

Any idea is all right in a first draft. If it's wrong, they can change it later. Many students are very anxious because they are asked to have right ideas immediately.

Why should people do freewriting if it's not going to be read?

It's the first step in getting kids to conceptualize better, to trust themselves, to write better. Teachers also. It reduces anxiety about writing.

What if we don't have time to use writing in our classes?

Time is the biggest bugbear, since curriculums often don't set aside enough of it for good writing programs. If they did allow it, though, learning would be more efficient in the long run. Cell biologist and educator Mary Bahns has observed that teachers who do not use writing do twice the amount of teaching as those who do. Even with little time to write, however, it can still be used often, and make a big difference.

Who is it that makes the decision to use writing to learn?

It would be good if school boards and principals did, but often teachers must do it on their own, and try to convince others that it is worthwhile. A good thing to advocate would be a special course that provides time for students to experiment with words and write freely.

Is writing really *that important to science and mathematics?*

Many of the brightest math and science students have enormous trouble making the transition from plugging numbers into formulas to understanding the science and math in higher levels of education and the professions. They fail because they can't relearn everything immediately. The loss to themselves and the society is tremendous—and unnecessary. Writing can help them avert this failure, because it enables them to approach their subjects the way real scientists and mathematicians approach them.

Is there anything that theater can contribute to science writing?

Certain techniques may be useful. When reading science writing aloud, it is good to know when you have the attention of the entire audience, why you are losing it if you don't, and how to get it back. It is good to know the different uses and values of listening, conversation, silence for thought, vocal projection, clear mental images, and more. Science can be as dramatic as a good performance, but this quality is often overlooked. All writing needs a genuine voice and the authenticity of one's voice is critical to good theater.

That's all well and good. I still want my kids to be exposed to . . .

Stop! Stop! Expose them all you want, but are you sure they are seeing? Are you sure they can tell you what they are seeing? Are the test-designers and the people at the top of educational bureaucracies listening to what the students are saying that they are seeing?

What if my classes are too big to use good writing techniques?

If you lose some time to reorganize the students into small, self-teaching groups, you will gain time later because they will assume responsibility for their own work. The incentive for working without supervision is that they will be dependent on each other to develop good writing.

How can science writing be more beautiful?

Often the beauty of many science texts comes as a surprise to students who have read predominantly textbooks. My approach is not to speak of writing as beautiful (or not), but to give many examples of great writing, then expect the students' writing to be more beautiful. Here is an example of beautiful writing (though in translation) from Leonardo da Vinci's notebooks:

What Sort of Thing the Moon Is

The moon is not of itself luminous, but is highly fitted to assimilate the character of light after the manner of a mirror, or of water, or of any other reflecting body; and it grows larger in the East and in the West, like the sun and the other planets. And the reason is that every luminous body looks larger in proportion as it is remote. . . . And if you could stand where the moon is, the sun would look to you as if it were reflected from all the sea that it illuminates by day; and the land amid the water would appear just like the dark spots that are on the moon, which, when looked at from our earth, appears to men the same as our earth would appear to any men who might dwell in the moon.

—Translated by Mrs. R. C. Bell

How can I make note-taking be more interesting for my students?

Students like to pass personal notes to each other, and one of the problems with class notes is that this element of communication is missing. Sometimes when looked at later, class notes can even lack clarity for the notetaker. Class notes can be valuable, but often students will immediately pick up their pens and begin to take notes when a lecture or discussion begins, even if you exhort them not to. For many students the taking of notes in the classroom is a way to avoid thinking.

Experiment with these methods: in some classes allow no pens or paper at all; in others allow notetaking; in others make it mandatory. In classes in which no notes have been taken, a summary writing exercise can work well at the end of class. Another possibility is to say "I'm going to have all the fun taking notes and you can't take them." It's interesting for the teacher to make notes while students are reading aloud and having discussions. The teacher begins to fade into the background and the students begin to manage the class.

Anthologies of class notes make good reading for all. Encourage your students to contribute *all* their notes, without censoring or selecting them.

Numerous kinds of notebooks can be made use of: dialectical notebooks to be shared with other students; journals of classroom activity; notebooks to be commented on by the teacher; notebooks that interrelate one subject with another; notebooks set aside for freewriting exercises; notebooks that contain only new ideas; notebooks set aside for summary writing or writing done outside the classroom, and so on.

How can mathematics papers become more lively and experimental?

Mathematics papers and textbooks, though well illustrated, often miss the mark for students. They seem dry and, of course, profuse with abstractions. On the one hand, it is useful to encourage students to become

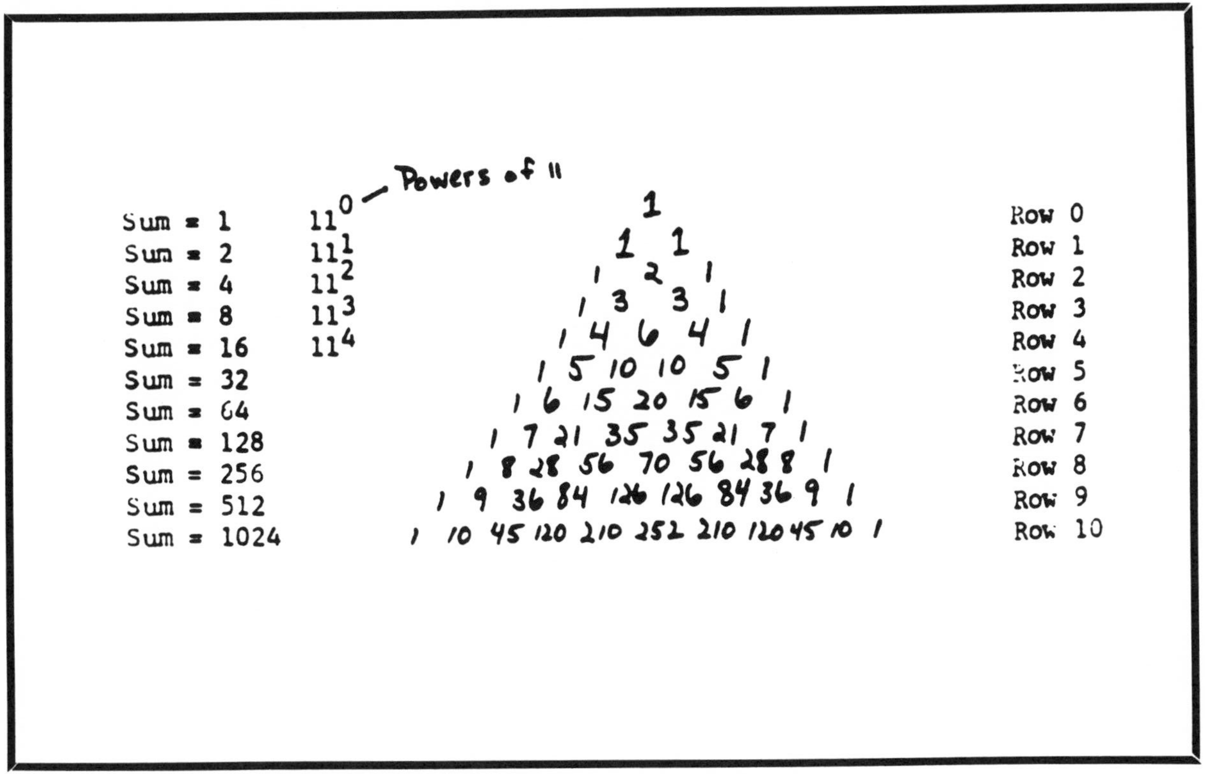

Fig. 2: Pascal's Triangle and some of its attributes.

even more abstract than ever. On the other hand, students can experiment with introducing ideas like Franklin White's piece below and with lighter, less formal tones and styles. It is useful and often compelling to write a math paper in the first person—to define the development of the concept as the author sees it, even to tell in what room, in what class it was first perceived, how other students reacted, and so on — the "surroundings" of the idea.

Here are two examples from papers on the Fibonacci numbers. The Fibonacci numbers are a sequence of integers in which each integer after the second is the sum of the two preceeding integers, as in 1,1,2,3,5,8,13,21, etc. The Fibonacci numbers are based on the breeding habits of rabbits and named for the thirteenth-century mathematician who discovered this math phenomenon.

> I'm not done yet. Fibonacci began thinking again. (This is getting dangerous!) What if he could apply his set of numbers (which he so egotistically named after himself instead of after those poor rabbits) to Pascal's Triangle? Are you wondering what Pascal's Triangle is? Wonder no more! I'm here to save your brain from exploding with wonderment! —*Rachel Fluty*

> It has been discovered that in plants there are spiral arrangements of seeds on the face of certain varieties of sunflowers. The numbers of spirals in the two sets are different and tend to be consecutive Fibonacci numbers. There has been a discovery of one mammoth sunflower with 144 and 233 spirals. —*Franklin White*

The French philosopher and mathematician Blaise Pascal invented "Pascal's Triangle" (see fig. 2). Here is the conclusion of a paper on Pascal's Triangle by Dean Austin:

> When you have a problem included within the triangle, you know that one way or the other through trial and error, you'll get to the right answer. This can also be used in everyday problems. You'll face your problems knowing that there is an infinite number of ways to solve them. You should list your problems and attack them through every corner until finally you're enlightened by a most convenient solution. One must have patience and flexibility of mind to get by in life, and the triangle, Pascal's Triangle, if I may say so, is a good beginning.

It would be worthwhile for teacher and students to write a math textbook together. Also, publishing a math newspaper or journal could become a good exercise in ingeniousness and humor.

Which science writing ideas will make the greatest impression on students?
Freewriting makes a great impression in school because it introduces

the non-scholastic notion of doing something that will never be graded or judged, an exercise of the mind done for its own sake.

Writing messages to be sent into outer space is popular, as is working with permutations and forms in writing, forms such as the sestina. Writing a history of one's own ideas, as Einstein did in his *Autobiographical Notes*, is an inspiring way to begin or end the school year or semester. The trading of papers among students for criticism and comment works out well, and the ultimate publication of the students' science writing in an anthology or magazine creates a terrific impression.

What about the frequency of the teaching of science writing?

For visiting writers who work with students once a week, it becomes obvious that students become twice as good at writing if you teach twice a week. If students write every day, the wonder of it increases more than proportionately, test scores improve and students' ability to enter the collegiate sphere with confidence and grace becomes greater.

Should philosophy be brought into the science curriculum?

Yes, including questions such as How is it possible to know anything? How can we distinguish truth from error, fact from prejudice or opinion? What is the most characteristic quality about our universe? Can everything be reduced to matter or mass and motion? Or, is mind or spirit the fundamental stuff out of which everything comes? Do our ideas about the world report faithfully the nature of that world or are they mere ideas? If so, how do we know? These and other questions can be incorporated into the activity of writing science. Philosophical questions can be a vacation from other forms of scientific investigation, but require rigorous thinking and a range of writing skills.

What about individual conferences with students doing science writing?

Writing conferences with individual students seem even more valuable in the teaching of science writing than in the teaching of poetry and fiction. Because science students are rarely used to thinking of their writing as having any literary value, it's wonderful to have the chance to go over the work line by line and sentence by sentence in one-to-one or small-group conferences, dealing with inspiration, new ideas, methods of expression, structure, humor, etc.

Why publish student writing?

For students, the prospect of publication puts all their writing, including science writing, in a different light: they become aware of an audience. It is difficult to fathom why so many school papers are seen only by teachers and maybe parents, not even by one's fellow students, much less by a wider audience.

Writing for publication in a magazine, journal, or book can often be the most valuable experiment to be made in the classroom. Writing in the schools for only one other person (the teacher) limits expansiveness and objectivity. Wild speculations, precious meanderings, beginning attempts, and succinct essays must be shared, at the very least by all the people in the class. In most schools it is possible to create mimeographed or xeroxed anthologies. High school students can take charge of the designing, typing, and collating.

Magazine Fundamentals is a useful guide to producing a school magazine. It is available from the Columbia Scholastic Press Association, Box 11, Central Mail Room, Columbia University, New York, N.Y. 11217.

How can writing fit into the forty-minute science class period?
Begin with freewriting for three to five minutes. A myriad of possibilities then ensues, among which are:

1. To divide the period among a presentation, another writing experiment, and the sharing of that experiment in a reading.

2. To just do everything fast, incorporating writing with small group discussions (seven students in each group), with a return to a five-minute summary writing.

3. To make use of dialectical notebooks (see the Writing Experiments section).

4. To devote one class entirely to a presentation and writing and the next class to the sharing of the writing and discussion.

5. To spend twenty minutes presenting and discussing, having half the students taking notes and the other half not, then sharing the results. Which half learned more? Variation: appoint only one notetaker.

6. To interrupt lecture and discussion with unexpected five-minute writing periods.

7. To ask some students to be notetakers during the talk and discussion, others to write discursively or in poem form afterwards.

8. Ask your students for ideas.

All too often, in the forty-minute period, by the time everyone has gotten warmed up, especially to writing, it's time to leave. I guess we have to adapt to this pace, but the feeling can be alleviated by being rigorous in terms of time on some days and loose and extending over into other classes on other days. Also, sometimes it's difficult to remember that one is still teaching when one is not talking, but just writing and thinking along with the students in the classroom.

Is there a danger in the controversial nature of science?

Sometimes in science writing, the most popular subjects among students have moral, ethical, religious, or political implications, for instance the question of evolution or current issues relating to human reproduction. Of course the danger for teachers is in loving the excitement of the controversialness at the expense of objectivity. Yet both students and teachers can part with their objectivity for a while and then regain it. All this makes the life of science exciting and full of risk.

Do we avoid "directed freewriting" because it seems like work?

The value of writing in the science classroom can be undercut by the desire to keep writing a pleasure and not make it a chore. Often I have chosen not to assign directed freewriting because the students said it seemed more like regular schoolwork than our other experiments. They're right, but the teacher's attitude should communicate a cheerfulness in the necessity for certain kinds of work.

What are some ways to approach a scientific problem through writing?

Besides notetaking, writing essays, and the ideas listed in the Writing Experiments section of this book, there are other possibilities: setting aside a particular time each day to think and write about a scientific question, perhaps at lunch time or before or after sleep or another of the day's meals; or, setting aside a problem for a month or two, then writing about it, then letting another month go by and writing about it again, assuming the mind is doing its work without overt thinking (though this might not be practical for students who have close deadlines, it is highly recommended by the philosopher and mathematician Bertrand Russell).

What if the teaching of science writing doesn't seem to be working?

If, in some classrooms, writing about science just doesn't seem to work, and the students are resisting, the thing to do is just get some writing going in the science classes anyway, writing unrelated to science. By creating a sense of pleasure in the process, you might, in the long run, get what you want.

Should teachers act more like students?

Yes, and the other way around. Both students and teachers should be less all-knowing, sit in each other's seats.

Should all student writing be corrected?

No. Notebooks and samples of freewriting, if they are perused at all, should definitely not be subjected to traditional correction on the basis of spelling and grammar. Also, exploratory work, notes, and investigations in writing in various forms should be left alone, so that students won't have to worry about questions of judgment or audience while writing in these ways. The main thing is to let words reflect thought, to let divagations occur, and to let the mind become interested in structure and transitions. Sometimes, writing that is intentionally non-grammatical can lead to new ideas. When it comes time to edit a final piece or essay, it's fun to let the students edit each other's papers and to see how they handle it.

Why are students so busy? Is there time for speculative work?

The speculative, the theoretical, the impractical, the risky, the uncertain: I think students who have eight forty-minute periods in a day have little time for this sort of learning or pondering.

As a poetry teacher who often goes from class to class rapidly, without time to think about what came before and what is to come, I find that this pace interferes with my own writing abilities. This aspect of the current American school is one of the reasons to make use of freewriting techniques, which provide peaceful and valuable thinking time in the course of a sometimes intellectually ravaged day.

What do you do when writing becomes boring?

It's good for students to know, at some point, that writing can be a tedious task. When writing and writers come into a school or class for the first time — or when teachers allow time for writing — the enthusiasm, break from routine, and exciting exercises introduced, along with the chances for class readings and publications, can lead to the idea that writing doesn't ever have to be hard work. There's a difference between "This is boring because I hate writing" and "This is boring because it's difficult." If difficulty becomes an issue, I've found it useful to have students work at home, where sometimes they can concentrate better. Not all students will do writing at home if grades aren't being given, but those who do will have fun. If boredom is the issue, it sometimes helps to bring a large number of science texts to class and leave them lying around for students to peruse. Also, younger students sometimes love what older students find boring, and vice versa.

Why do so many schools ignore writing as a tool for expression and the understanding of concepts?

I don't understand it. When my daughter Marie was in the fifth grade, I asked her teacher why Marie was considered to be a good student when she couldn't really write, though she read a lot. The teacher replied that that was the way it was nowadays. Later I asked an eighth-grade teacher in a school where I was working why writing could not be a five-day-a-week class in the school, even if it could not yet be introduced "across the curriculum." Her answer was that there was no teacher qualified to teach it and that, if there were, that person wouldn't want to have so many papers to correct. A twelfth-grade student who had attended schools in three foreign countries told me, "In those schools, all we ever did was write, all day long."

I don't know the whole answer to this question, but I do know that memorizing facts and getting good test results seem to have supplanted reasoning and writing in many American schools, perhaps because of the increasingly job-oriented nature of the American school.

Will this writing help improve standardized test scores?

That's a red herring, but writing probably will help, if used with discretion. The point is that such tests often preclude the effective learning implied by the use of good writing. "Teach for the test" is the advice of administrators, when they should be saying "Make sure the students learn."

Can computers be used effectively in science writing?

We refer you to the following outline of a presentation by Ruth von Blum at Bard College's Institute for Writing and Thinking's 1987 conference on the Role of Writing in Learning Mathematics and Science. Ms. von Blum is Director of Science, Education Systems Technology Corporation in San Diego, California, and designer of a computerized writer's aid for students in English composition called WANDAH, currently marketed as HBJ Writer.

ROLE OF THE COMPUTER IN STIMULATING SCIENCE STUDENTS' WRITING/THINKING

One of the major goals of science instruction is to have students get involved in "scientific process." This means observing, making hypotheses, predicting, setting up experiments, observing what happens, figuring out what results mean, drawing conclusions. It also means taking risks, playing around with ideas, not being afraid of being wrong, and working with others. The computer can help the students think and write about science at every stage of instruction.

A. *Pre-experiment*

Often, coming up with a good, testable hypothesis is the hardest part of a science experiment (ask any graduate student!).

1. Scientific "brainstorming"

The computer can help students conceptualize a problem and clarify a question. An "idea processor" (like the one I have used to write this outline) can be very useful to stimulate and organize ideas. It lets the student get ideas down, expand them, move them around, organize them. These programs (like Thinktank, MORE, Maxthink) force the student to translate fuzzy ideas into concrete words.

2. "Invisible writing"

It sounds weird, but it works! When students are stuck, and thinking and writing are blocked, it sometimes helps them to "turn off the screen" and write without seeing what they write. It forces them to keep their ideas at the front of their consciousness, to move ahead and not to go back and revise endlessly.

3. Lesson-specific aids

Even if the program cannot analyze what they write, students can be given hints based on the specific topics they are dealing with to help them formulate their ideas.

4. Summary statements of hypothesis made by student

Students can write a summary statement of the hypothesis that they are testing and its rationale. This can be available as an on-line reference for the student as she is conducting her experiment. (I will give a brief aside here about my Berkeley biology class for majors where over sixty percent of "independent investigations" done by the students had *no* hypothesis, either stated or implied.)

B. *Experiment*

Of course, the computer can provide the experimental environment for the student to work. But, it can also offer tools to writing and thinking as experiments are conducted.

1. Pop-up notebook for taking notes during experiment

One of the most useful tools is a pop-up notebook. It lets the student take notes on the experiment as it is being conducted. It also lets the student transfer charts and graphs from the experiment on screen to the notebook. Having an "electronic notebook" not only saves time, but it keeps attention focused on the screen, and is inherently more motivating than an off-line notebook. Moreover, the computer can offer help on note-taking skills.

2. Experiment-specific helps to data collection

If the program knows what hypotheses the students are testing, it can provide specific hints and helps, including what to watch for, and think about/write about as the experiments progress.

C. *Organization and analysis of data*

Of course, the computer can help the students organize and analyze their data. What's more, it can help them write about (think about) the data as they organize it.

1. Graphing and charting data

On-line graphing utilities help students see data in different ways. Having students put in words what is happening on the graph is a good way of seeing if the student really understands what the graph or chart says.

2. On-line notebook

The on-line notebook lets the student analyze directly what is on the screen.

3. Specific helps

Once again, the program can help the student specifically analyze the results by offering probing questions and hints.

D. *Presentation of results and conclusions*

Drawing conclusions from the data is the last stage in the writing/thinking process. And the computer can help here as well.

1. Word processor — makes writing reports easier

Everything that the student puts in her notebook can be transferred to the word processor. This saves much time and effort.

a) Helps ideas flow

Because corrections can be made so easily, and text can be moved around, writing with the word processor (once you get used to it) can really help ideas flow.

b) Facilitates revision

Using the word processor also means that students can revise their work more than they would if writing by hand or on a typewriter. If these revisions are on a structural (and not just a surface) level, they involve re-thinking and not just moving words around.

E. *Critical review of the experiment*

Another place where the computer can help the student to think/write about her experiment is in the critical review stage. Here the computer can provide some content and style review.

1. Peer analysis

Writing on-line makes on-line review by peers (and even teachers) easier. The reviewer can comment directly in a students' report, without damaging the integrity of the report. (The reviewer's comments are stored in a copy of the student's report in a separate file.)

On-line guides can help the reviewer look for common flaws in logic or presentation.

2. Student review of mock reports

Reviewing another student's work benefits the reviewer as well as the student who is reviewed. The computer can provide "mock reports" for review, and can give specific help to the student doing the reviewing. The program "knows" where the defects in the report are, and can help the student think, and write, about what is wrong and why.

3. Textual review

In addition to peer review, the computer can help the student revise reports in several ways.

a) Revising for organization

The computer brings each paragraph in a report onto the screen and asks the student to pull out the sentence with the main idea from each paragraph. Looking at such an ex post facto outline is very helpful for seeing the logical structure (or lack thereof) in a report.

b) Style

Since the computer is so good at counting things, it makes a great quantitative style checker. Students can examine different writing styles and their appropriateness to different audiences.

Sentence length graph

This is very useful for discovering uniformity (and often monotony) in the prose.

To-be verbs

Looking at this points out passive construction as well as overuse of the verb "to be."

Nominalizations

Using lots of nominalizations (verbs turned into nouns) makes for rather static writing. (For example: "The action was undertaken by the author," instead of "I did it.")

Transition words

The absence of transition words can point out choppy writing.

Thesaurus

An on-line thesaurus can help with word choice.

If requested, the program can look for standard scientific report format

c) Mechanics

Spelling-checker

Commonly misused words/phrases
The computer can point out words or phrases that are commonly mis-used, and the student can decide if they are used properly in the writing in question.

Samples

The following samples from the literature of science and mathematics by no means represent a thorough survey of the field. If you are developing essays along the lines described by the Essay Development Workshop section, you may find material for the "Inspiration" phase here. Each example is chosen on the basis of its ability to inspire or demonstrate an interesting point about writing. Many contain provocative ideas that may stimulate good classroom discussions. In most cases, the passages are excerpts from larger works, and can only suggest the depth and subtlety of their sources. They are intended here to ignite interest, not to explore subjects.

Some of the samples are taken from a collection edited by Martin Gardner, *The Sacred Beetle and Other Great Essays in Science*, a good resource for any science writing program.

Rachel Fluty

Rachel Fluty's essay is one of a series written voluntarily by members of the Manhattan Center for Science and Mathematics math club. A combination of research, good math, and personal enthusiasm, her essay breathes life into a subject that unfortunately is often presented too dryly to be appreciated. Rachel was sixteen when she wrote this piece.

Fibonacci Numbers

The Fibonacci Numbers are a series of numbers that were originated by a very intelligent man named Leonardo Fibonacci (or Bonacci or dePisa) when he proposed a mathematical puzzle. He decided to analyze the breeding of rabbits. He noticed that rabbits were very active and multiplied rapidly. After watching them for some time, he realized that when rabbits were a month old they were already able to reproduce. Given the chance that two lovesick rabbits met, there definitely would be a baby bunny or two the next month. Fibonacci decided he would breed rabbits for one whole year and see if a pattern existed at the end of the year.

Here is how Fibonacci approached the problem: he set up a chart consisting of four columns. The first column contains the number of pairs of breeding rabbits at the beginning of the given month (as you'll see, it is to your left on the chart). The second column contains the number of pairs of nonbreeding rabbits at the beginning of the month. The third column contains the number of pairs of rabbits bred during the month. Finally, the fourth column contains the number of pairs of rabbits living at the end of the month. (Have I lost you yet? If I haven't, good. There's still more to come. Fibonacci never gives up!) After he set up the chart, he deduced that any term is the sum of the two immediately preceding terms.

In other words (or English):
$$U_{n+1} = U_n + U_{n-1} \ (\text{if } U_0 = 0 \text{ and } U_1 = 1, \ldots)$$
For example, where n = 2:
$$U_{2+1} = U_2 + U_{2-1}$$
$$U_3 = U_2 + U_1$$
$$3 = 2 + 1$$
$$3 = 3$$

IT WORKS!!!!!

And now, the moment you've all been waiting for:

Fibonacci's Chart for Breeding Rabbits!

Month	1	2	3	4
January	0	1	0	1
February	1	0	1	2
March	1	1	1	3
April	2	1	2	5
May	3	2	3	8
June	5	3	5	13
July	8	5	8	21
August	13	8	13	34
September	21	13	21	55
October	34	21	34	89
November	55	34	55	144
December	89	55	89	233

I found this puzzle very interesting and original. This just goes to show the wonder of the world of numbers and the infinite amount of ways you can use them.

I'm not done yet. Fibonacci began thinking again. (This is getting dangerous!) What if he could apply his set of numbers (which he so egotistically named after himself instead of after those poor rabbits) to Pascal's

Triangle? Are you wondering what Pascal's Triangle is? Wonder no more! I'm here to save your brain from exploding with wonderment!

Pascal's Triangle is formed by beginning with the number 1 at the apex of a triangle in which n = 0. All the other numbers are sums of the two numbers just above them (for example: $10 = 4 + 6$). Each of the rows is the coefficient of any binomial expression raised to the nth power; for example:
$$(x + y)^3 = 1x^3 + 3x^2y + 3xy^2 + 1y^3.$$

n = 0						1						$(x + y)^0$
n = 1					1		1					$(x + y)^1$
n = 2				1		2		1				$(x + y)^2$
n = 3			1		3		3		1			$(x + y)^3$
n = 4		1		4		6		4		1		$(x + y)^4$
n = 5	1		5		10		10		5		1	$(x + y)^5$
n = 6	1	6		15		20		15		6	1	$(x + y)^6$

"N" is equal to the power of the binomial expression. This triangle could expand until infinity! Fibonacci adapted the Chinese triangle, a modified version of Pascal's. By displacing the numbers of Pascal's Triangle to the left, you get the Chinese Triangle.

		r	0	1	2	3	4	5	6
Fibonacci	n		0	1	1	2	3	5	8
Series	0		1						
	1		1	1					
	2		1	2	1				
	3		1	3	3	1			
	4		1	4	6	4	1		
	5		1	5	10	10	5	1	
	6		1	6	15	20	15	6	1

If you add any array diagonally, it will correspond to a Fibonacci number. You can also get a Fibonacci number by a summation (Σ) such as:
$$_nC_0 + _{n-1}C_1 + _{n-2}C_2 + _{n-3}C_3 + \ldots + _{n-r}C_r \text{ or}$$

$$\sum_{r=0}^{n} {}_{n-r}C_r$$

The series terminates when $n - r = 1$ or 0.

For example: let $n = 9$.
$$_9C_0 + _8C_1 + _7C_2 + _6C_3 + _5C_4 = 1 + 8 + 21 + 20 + 5 = 55 = U_{11}.$$
If you do this in another direction you get:
$$_9C_0 + _8C_1 + _7C_2 + _6C_3 + _5C_4 = _9C_9 + _8C_7 + _7C_5 + _6C_3 + _5C_2 = U_{11}$$
Let me share a secret with you. Fibonacci was very intelligent and had

a creative mind. If you just stopped yourself for a minute and contemplated these problems or any other problems in math, wouldn't you be creative also? It's not hard to think this way. Fibonacci did it and look at what he accomplished! Just imagine all the amazing things math has done: it shelters you and it clothes you. Without math, I see no possible way of surviving in life. All you have to do is think about it for a while. I hope you enjoyed my story.

•

Tim Murphy

Tim Murphy was a junior at North Andover High School, North Andover, Massachusetts, when his science fiction parody won the 1986-87 Edison Electric Institute's science fiction short story contest.

What It Will Be Like to Live in the Future
"Boy, oh, boy!" I say to myself as I rapturously shut off the television set in my bedroom. "It sure is great watching 'The Jetsons'—all those ionic power pellets, and girls with hairdos like 1930's art deco lampshades, and amusement park rides in your very own home and office! I sure bet it would be great to live in the future...all fun and games, for sure!"

And with that, I crawl back to bed for some much-needed sleep—afternoon sitcoms *do*, after all, require a great deal of mental. "Ouch!" I cry, as my toe hits something on the floor. I look down to see my sister's beat-up Barbie Doll lying there, her movable plastic limbs distorted into the most bizarre positions, her thatch of limp, blond hair fanned out over her perfectly proportioned face. "If only Susie would keep her stupid dolls out of my room," I mutter, as I fling the doll across the room near the heater. I flop on the bed and fall fast asleep, visions of Judy Jetson dancing in my head.

I woke up shortly afterwards. Feeling a bright light beating down on my tightly shut eyes, I opened them and peeped around—I was no longer in my bedroom, but in the thicket of a bush in the sunny forest. The strange thing, however, was that the bush was not made of wood, but a strange green rubber material, from the branches to the smallest leaves. The same was true of all the foliage, and I noticed that even the squirrels and birds moved in such a stilted manner that they seemed to be taken from some animated children's TV special. On closer inspection, I discovered—to my amazement—that they were made of a strange plastic substance, too!

I heard a noise from within the depths of the forest and hid behind the bush, peering silently over the edge. My confusion turned to awestruck fear as a large group of people emerged from the foliage into the clearing, all of them babbling in the same robotic monotone. I squinted and was shocked to realize that, although they resembled human beings from a distance, they were not real people at all — but merely walking, talking mannequins with a waxy plastic skin, eyes that did not move but were mere decals, and limbs that twisted in 360 degree circles in their sockets. The lead mannequin, a shapely blond in a spandex jumpsuit and stiletto heels, barked in an imposing computerized voice to the others: "Onwards! We must find Judy before the sun gets too bright and we begin to decompose."

Before I had a chance to wonder "Who's Judy?" I was scared to pieces to feel someone tapping on my shoulder. I spun around frantically—only to see, crouched behind me in the thicket, Judy Jetson herself! There she stood, in her radioactive ponytail and twenty-first century go-go boots— but human at least — and I didn't hesitate to speak:

"Judy Jetson! It certainly is good to see you, but I thought you were only a cartoon character."

"Don't be silly, I've been existing all along in the twenty-first century. One night in 1965 some executive producer at Hanna-Barbera fell asleep and entered the next dimension in one of his dreams. When he woke up, he had this great idea for a cartoon about a twenty-first century family, but he never knew where it came from."

"Oh, I see—that's simple enough. But who are these hideous polyurethane creatures that have just come through the forest?"

"Oh, it's a long ugly story, but I'll tell you. They are aliens from the planet Plastica in a galaxy far, far away. Not too long ago, they invaded the Earth. Unlike humans, they do not spit saliva, but silly putty. When this silly putty comes into contact with any human creature, the creature turns into lifeless plastic. So far, they have transformed every organism on this planet except me. They realize this and have been searching for me through this forest for days. I've been living off Fluffernutter and Screaming Yellow Zonkers for what seems like years now. And I don't intend to let them get me, either! I have a lifetime supply of Oil of Olay left and it would be useless if they turned me into some plastic dummy! Really, you've got to help me."

"Gee, Judy. . . is there any way you know of to destroy the plasticoids?"

"Yes, if exposed to high temperature they will melt on the spot and never reconstruct. That's why I get a break from hiding between eleven and one o'clock, because they go underground. At least it gives me a chance to work on my tan. . .peak tanning hours, you know. God, I have such a headache."

She pulled an aspirin from her purse and popped it into her mouth. Then she hesitated, grumbled, "This isn't half the strength I need for this migraine," and dropped it into a small tube of murky blue substance. Instantly, the aspirin expanded to the size of a potato.

"That's better," she said, and took a bite of the inflated tablet. Suddenly, I got an idea.

"Judy, let's go back to your house—I think I know a way to stop the plasticoids."

She programmed her computerized jet pack to take her home, affectionately grabbed my hand, and soon we were airborn for her house, speeding through the hot summer skies for the slick, modular dwelling that I so eagerly entered each afternoon at four o'clock. It stood as I expected, completely equipped with wall-to-wall computer terminals, robots, and an audiovisual circuit to communicate with the Jetsons' relatives on Mars and Venus.

"Now," I said as she offered me my choice of a sushi or a quiche lorraine food tablet, "What we need is a microwave oven—you know, one of those new things with all the digital buttons on them."

"Isn't that neat—we were reading about them in history class just the day before my professor was turned into a Ken Doll. I brought in an antique version that belonged to my great-great-great-grandmother for a class oral presentation. I'll go get it."

We flew back to the clearing in the rubber forest, decrepit old microwave in tow. I set the oven smack in the middle of the clearing.

"We need some of that murky blue stuff that you use to enlarge the aspirin," I said. She gave me a vial of it and I poured the entire thing over the microwave. Instantly, it expanded to the size of a truck. I heaved open the oven's mammoth door, and gave instructions to Judy: "The plasticoids should be reemerging from their subterranean sanctuary now. You go bait them to the clearing, then come hide behind the bush with me."

Judy stalked brazenly to the center of the clearing and shouted, "Here I am, you bumbling eraser-heads! Come and get me!" and then rushed behind the bushes.

Instantly, the whole flock of plastic aliens came marching into the field, aggravated blips and electronic obscenities pouring from their motionless, moldenshut lips. The lead plasticoid spotted the gargantuan microwave oven and said, "It must be a special spacecraft sent to us from Plastica to lead us to Judy! File in, troops."

As soon as they had crammed themselves into the oven, I rushed forth and slammed the door shut. Judy emerged, saying, "Allow me the honor, please. I've never tinkered with these electronic doodads before." She pressed the "MAXIMUM ENERGY" level and we both watched through the glass as the synthetic fiber of the plasticoids' hair was singed to ashes. Soon enough, their "skin" began to bubble, then drip to the floor, then form a pool of liquid plastic. A few minutes later there were no more aliens, just a pancake-like disc of cooled-off plastic at the bottom of the oven.

Judy pulled out another vial of orange substance and poured it on the microwave, instantly reducing it to the size of a matchbox. "So much for historical value," she sighed, as we buried the tiny thing in the dirt, saying good-bye to the nasty plasticoids forever.

"But what about all the native Earth organisms that are still in plastic form?" I asked.

"Don't worry. They must receive daily doses of the alien's silly putty saliva to stay in plastic form, so all will be back to their original physical and chemical state within 24 hours." Then she batted her psychedelic eyelashes at me and cooed, "How can I ever repay you? You saved my life, not to mention the most successful sci-fi cartoon ever and God-knows-how-many possible spin-offs and promotional products."

I knew just the way. . . .

I awoke, sweaty and disoriented on my bed, just a moment or two later.

"Darn," I said. "Just a dream, but why does it always have to end at the best part?" Then my nostrils were assaulted with the odor of burning rubber. I turned to the heater by the wall of my room and noticed that what had once been a Barbie Doll was now a pool of hot rubber. One tiny, plastic stiletto-heeled shoe protruded gruesomely from the muck.

"No matter," I said. I mopped up the mess and trotted over to the TV set. It was time for "Lost in Space"—and more down-to-earth, scientifically correct accounts of what it will be like to live and love and fight evil villains in the twenty-first century.

But that Judy was one "far-out" ticket!

•

Jane Goodall

When it was published in 1971, primatologist Jane Goodall's In the Shadow of Man *became an instant classic in the field of primate research, and remains today an example of the power of science to enthrall. In an appendix, Goodall poses the question "In what ways is the study of chimpanzee behavior relative to the study of man?" and answers it, in part: "The chimpanzee is our closest living relative. Recent research in biochemistry has shown that in some ways the chimpanzee is as close to man as he is to the gorilla. Neuroanatomists have pointed to the fact that the circuitry of the chimpanzee brain resembles the circuitry of the human brain more closely than does that of any other species. . . . The methods by which we raise our children would be a point of interest to all. Our studies at Gombe (in Tanzania, Africa) on different mothering techniques and on the stages of infant development have already proved of interest to child psychologists and psychiatrists. Our work highlights the importance of the mother-offspring affectionate bond in chimpanzees but we are not sure why some adult chimpanzees maintain closer bonds with their families than others. The answer may be significant to the understanding of some human family problems." The following passage is from the chapter called "The Infant."*

The Infant

The birth of a baby is something of an event in many animal and human societies. In the chimpanzee community, where mothers only have an infant about once every three and a half to five years, births are relatively few—not more than one or two a year in our group of thirty to forty individuals. So the appearance of a mother with a brand-new baby often stimulates much interest among the other chimps.

When Goblin was first introduced to a large group, he was two days old. He could only grip on for a couple of steps without support from his mother, Melissa, and he was still attached by the umbilical cord to the placenta. The group was peacefully grooming, but when Melissa approached the tree and began to climb it I sensed the tension as first one and then another chimpanzee stared toward the new mother. Fifi instantly swung over toward her. Melissa, moving carefully, went to greet Mike, pant-grunting submissively and reaching to touch his side. She presented and he patted her rump, but when he moved forward, staring at Goblin, she hastily moved away. It was the same when she greeted Goliath. He too wanted to see the baby more closely. So did David and so did Rodolf.

After five minutes Mike began to display, leaping through the trees and swaying the branches. Melissa, screaming, jumped away from him. As she did so, the placenta almost caught in some twigs. I was suddenly concerned that the newborn might be torn from his mother's breast, but Melissa gathered up the cord just as Goliath displayed toward her. Soon the entire group was in confusion — all the adult males leaping around and swaying branches, Melissa and also the other females and youngsters rushing out of the way and screaming. It looked for all the world like a wild greeting ceremony for the new baby, though in reality it was undoubtedly provoked by a sense of frustrated curiosity: Melissa simply would not let the males get close enough to examine the baby properly.

Eventually things calmed down and the males began to groom each other again. Then Fifi and the other young females in the group gathered around the mother and baby to stare long at Goblin. If they moved too close, Melissa threatened them with a soft bark or upraised arm, but she did not move away and they were able to look their fill.

Since that day we have seen similar displays when new mothers first appeared with their babies, though normally only when the mothers themselves were young. There is usually far less commotion when an old female, such as Flo, appears with yet another infant. Since the experienced female does not run away from them, the other chimpanzees are able to satisfy their curiosity. We have seen a group of four males sitting calmly very close to an old mother and staring fixedly at her newborn. When the mother runs away the situation is potentially very dangerous for a newborn baby, since for the first few days of its life the chimpanzee infant often does not seem able to grip well to its mother's hair. Usually there is the added danger of a dangling placenta, for, so far as we know, the chimpanzee mother in the wild makes no attempt to break the umbilical cord herself. We have never actually seen an infant dropped or hurt during these wild displays. On the other hand, we know of several babies that have mysteriously disappeared during the first few days of life.

During the six years since Flint and Goblin were born, twelve healthy infants have been born to our group, and although several died before they were a year old our observations on them and their mothers have taught us much. Babies less than five months of age are normally protected by their mothers from all contact with other chimpanzees except their own siblings. Infants from the age of three months onward often reach out to other chimps sitting nearby, but usually their mothers pull their hands quickly away. It was, however, very different for Pom, one of the first female infants born into our group. Her mother, Passion, actually laid the baby on the ground the very first day of her life and allowed two young females to touch and even groom her as she lay there. But then, in all respects, Passion was a somewhat unnatural mother.

She was no youngster, this Passion. She had been fully mature when I first came to know her back in 1961, and I know she lost one infant before Pom's birth in 1965. If her treatment of Pom was anything to go by, I suspect Passion had lost other infants, too, for Pom had to fight for her survival right from the start. When she was a mere two months old, she began to ride on her mother's back — three or four months earlier than other infants. It started when Pom hurt her foot badly. She could not grip properly, and Passion, rather than constantly support the infant with one hand as most mothers would have done, probably pushed Pom up onto her back. The very first day that Pom adopted this new riding position, Passion hurried for about thirty yards in order to greet a group of adult males, seemingly quite without concern for her infant. Pom, clinging on frantically, managed to stay aboard — though much older infants, when they start riding on their mother's back, usually slide down if their mothers make sudden movements.

We expected that when her foot was better Pom would revert to clinging underneath Passion. Nothing of the sort happened. It is probably more comfortable for the mother when the child rides on her back, and having achieved this happy state of affairs three months early, Passion had no intention of letting things change. Even when it poured with rain and Pom whimpered as she tried to wriggle under her mother's warm body for protection, Passion seldom relented, instead again and again pushing her infant back on top.

Most of the mother chimpanzees we have watched were helpful when their small babies nuzzled about searching for a nipple. Flo had normally supported Flint so that suckling was easy for him throughout a feed, even when he was six months old. Melissa also had tried, though often she had bungled things and held Goblin too high so that his searching lips nuzzled through the hair of her shoulders or neck. But Passion usually ignored Pom's whimpers completely; if she couldn't find the nipple by herself it was just bad luck. If Pom happened to be suckling when Passion wanted to move off, she seldom waited until the infant had finished her meal; she just got up and went, and Pom, clinging for once under her mother, struggled to keep the nipple in her mouth as long as she could before she was relentlessly pushed up onto Passion's back. As a result of her mother's lack of solicitude, Pom seldom managed to suckle for more than two minutes at a time before she was interrupted by Passion, and often it was much less. Most infants during their first years suckle for about three minutes once an hour. Pom probably made up for her shorter feeds by suckling more frequently.

It was the same story when Pom started to walk. Flo, it will be remembered, was very solicitous when Flint was finding his feet, gathering him up if he fell and often supporting him with one hand as he wobbled along. Melissa was less concerned, and if Goblin fell and cried, merely reached her hand toward him while he struggled to his feet. Passion was positively callous. One day, previous to which Pom had never been seen to totter on her own for more than two yards, Passion suddenly got up and walked away from her infant. Pom, struggling to follow and falling continually, whimpered louder each time, and finally her mother returned and shoved the infant onto her back. This happened repeatedly. As Pom learned to walk better, Passion did not even bother to return when the infant cried — she just waited for her to catch up by herself.

When Pom was a year old it was a common sight to see Passion walking along followed by a whimpering infant who was frantically trying to catch up and climb aboard her moving transport. It was not really surprising that, during her second year, when most infants wander about happily quite far away from their mothers, Pom usually sat or played very close to Passion. For months on end she actually held tightly on to Passion with one hand during her games with Flint and Goblin and the other infants. Obviously she was terrified of being left behind.

•

John Allen Paulos

In the "My Turn" column of Newsweek, *John Allen Paulos, a mathematics professor at Temple University and the author of* Mathmatics and Humor, *identifies a new disease: innumeracy.*

Orders of Magnitude
Quick: how fast does human hair grow in miles per hour? What is the volume of all the human blood in the world? If you don't know, it's no surprise; even math students sometimes don't, either. The answers are perhaps intriguing: hair grows a little faster than 10 to the minus-8th miles an hour (that is, a decimal point followed by 7 zeros and a 1); the totality of human blood will fill a cube 800 feet on a side, or cover New York's Central Park to a depth of about 20 feet.

What does surprise me, however, is how often I find adults who have no ideas of easily imagined numbers: the population of the United States, say, or the approximate distance from the East Coast to the West. Many otherwise sophisticated people have no feel for magnitude, no grasp of large numbers like the federal deficit or small probabilities like the chances of ingesting cyanide-laced painkillers. This disability, which the computer scientist Douglas Hofstadter calls innumeracy, is so widespread that it can lead to bad public policies, poor personal decisions — even a susceptibility to pseudoscience.

Without some intuition for common numbers, it's hard to react with the proper skepticism to terrifying reports of the exotic disease of the month. It's impossible to respond with the proper sobriety to a warhead carrying a megaton of explosive power — the equivalent of a million tons of TNT. Without some feel for probability, car accidents appear to be a relatively minor problem of local travel while being killed by terrorists looms as a major risk of international travel. While 28 million Americans traveled abroad in 1985, 39 Americans were killed by terrorists that year, a bad year — 1 chance in 700,000. Compare that with the annual rates for other modes of travel within the United States: 1 chance in 96,000 of dying in a bicycle crash; 1 chance in 37,000 of drowning and 1 chance in only 5,300 of dying in an automobile accident.

There's a joke I like that's marginally relevant. An old married couple in their 90s contact a divorce lawyer. He pleads with them to stay together. "Why get divorced now after 70 years of marriage? Why not last it out? Why now?" Finally, the little old lady responds in a creaky voice, "We wanted to wait until the children were dead." A feeling for which magnitudes are appropriate for various contexts is essential to getting the joke.

It might seem that one way to combat innumeracy is for newspapers to use scientific notation — 9.3×10^7 instead of 93 million or 93,000,000; 2.2×10^{13} instead of 22 trillion or 22,000,000,000,000. In a sense we do this when we use the Richter scale for earthquakes, and decibels for sound intensity. But how many people remember that a 6 on the Richter scale indicates an earthquake 10 times as severe as a 5?

The media could also discuss probabilities in contexts besides weather forecasting — particularly in medical reporting. This would acclimate people to the concepts involved. Even more helpful when dealing with magnitude would be frequent comparisons that invoke everyday life. For example: knowing that there are approximately a million seconds in 12 days gives one a new grasp on that number. By contrast, it takes almost 32 years for a billion seconds to tick away. And the human species is not much more than a trillion seconds old.

Another quiz: how many trees were cut down last year to print books, articles, and newspaper columns on astrology and psychic readings? Whatever the number, its sheer magnitude leads me to an unappreciated consequence of innumeracy: a predisposition to believe in pseudoscience. Consider first an intriguing result in probability—a subject, by the way, that should be taught to everyone. Since a year has 366 days (counting February 29), there would have to be 367 people gathered together for one to be 100 percent certain that at least two people have the same birthday. How many people would be required for one to be just 50 percent certain? A reasonable guess might be 184, about half of 367. The unexpected answer: there need be only 23! For one to be 50 percent certain that somebody has a particular birthday, however—say, July 4—a crowd of 253 would be necessary.

The example may be odd, but the principle is quite general. That *some* unlikely event will come to pass is likely; that a *particular* one will is not. The pronouncements of psychics or astrologers are sufficiently vague so that the probability that *some* prediction will occur is very high. It's the *particular* incidents that are seldom true. Yet the overwhelming majority of incorrect predictions are conveniently forgotten and the correct ones are greatly magnified by publicity. Without a feel for number and chance, people can easily be misled.

The primary reason innumeracy is so pernicious is the ease with which numbers are invoked to bludgeon the innumerate into dumb acquiescence. Even though mathematics deals with certainties, its applications are only as good as the underlying assumptions, simplifications, and estimations that go into them. Any bit of nonsense, for example, can be computerized—astrology, biorhythmns, certain items in the military budget—but that doesn't make the nonsense more valid. Statistical projections are invoked so thoughtlessly that it wouldn't be surprising to see someday that the projected waiting period for an abortion is a year.

When I write of innumeracy, I'm not referring to ignorance of any abstract higher mathematics. I'm bemoaning a lack much more basic. Somehow, too many Americans escape education in mathematics with only the haziest feel for numbers and probability and for the ways in which these notions are essential to understanding a complex world.

•

William Carlos Williams

William Carlos Williams (1883-1963) espoused a poetic theory of "no ideas but in things." In other words, the meaning of a poem should emerge from the poet's clear-eyed presentation of the objective world. This objective view gave many of Williams's poems a refreshing clarity, which made his whimsicality all the more outstanding. In the following poem, Williams, who was a doctor, combines scientific rigor, dailiness, orneriness, and good humor.

Le Médecin Malgré Lui
Oh I suppose I should
wash the walls of my office
polish the rust from
my instruments and keep them
definitely in order
build shelves in the laboratory
empty out the old stains
clean the bottles
and refill them, buy
another lens, put
my journals on edge instead of
letting them lie flat
in heaps—then begin
ten years back and
gradually
read them to date
cataloguing important
articles for ready reference.
I suppose I should
read the new books.
If to this I added
a bill at the tailor's
and at the cleaner's
grew a decent beard
and cultivated a look
of importance—
Who can tell? I might be
a credit to my Lady Happiness
and never think anything
but a white thought!

•

Aurika Checinska

Aurika Checinska was a winner of the 1988 Westinghouse Science Talent Search contest. A junior at Stuyvesant High School in New York, Ms. Checinska combines her interest in physics and music in this excerpt from her paper.

A Three-Dimensional Comparison of the Sound Fields Generated by a Violin Using Various Bows and Various Bow Strokes

Abstract
In an anechoic chamber the decibel level of 16 frequencies produced on a violin by five different bows was measured and compared during the use of legato and spiccato bowing. The microphone was placed in five different positions. The statistical analysis described the sound field surrounding the violin and illustrated that the change of bow stroke did not affect the field. The comparison of the fields created by each individual bow demonstrated that the structural properties of a bow have an influence upon sound production. Knowledge of the directional output of the frequencies on a violin may be used to improve the acoustical properties of concert halls, the microphone set-up necessary to generate good recordings, and the overall performance of a musician.

Introduction
Although many books have been written about the development of the viol family, and consequently about the violin itself, research concerning the violin bow has begun only recently, thus the art and science of bow-making have not been as well explored. Whereas two centuries ago, the construction of the violin bow was limited to the familial secrets passed on and improved upon from generation to generation by such families as the Tourtes in France, the Dodds in England, the Nurnbergers in Germany, and others, the contemporary knowledge of bow synthesis approached a level of scientific preciseness when it began to involve mathematical and physical principles, as illustrated by Vuillaume and later expounded upon by others such as John C. Schelleng and Andreas Askenfelt.

For the performer, only within the last two hundred years has the bow achieved equal status with the instrument, as a primary vibrating factor in tone production. So little attention has been given to perfecting the bow itself as compared to the instrument, the expertise of which matured almost two centuries previously through the work of the Amati family, or the bow technique involved in performance, that the earliest development of the bow as used for playing stringed instruments is shrouded in mystery. Both historians and musicologists, for example Joseph Roda[1], David Reck[2], and Emanuel Winternitz[3] are presently devoting a great deal of time and effort towards researching the designs and development of the bow. For the earliest stages much speculation is involved, especially with regard to the geographic sources.

The most widely accepted theory states that the bow and the instruments were both derived from the hunter's bow.[4] Nevertheless, the earliest recorded use of the bow is found in sculptures and art works dating from about A.D. 1000. These first representations are quite primitive, illustrating a tool that appears to be made of a branch or pliable reed held in a convex shape by means of threads, strings, or even horsehair, tied at each end. In this simplest form, it was up to the player to control the tension of the hair with his hand as he played.[5]

It was not until the Renaissance that the bow was first considered a more common appliance. The emblematic literature of the 16th and 17th centuries attributed the creation of the bow to the ancient Greek poetess Sappho,[6] and many paintings of that era allegorically represented musicians with various types of viols and bows.

Commencing with the year 1620, a definite record of bow construction and development exists.[7] The first such bow is labelled "Mersenne, 1620" [fig. 3] and may be described as a flat stick which extends into a tapered and convex-curved 'head' with a fixed horn-shaped 'handle' (according to present day nomenclature — the frog). By 1680 a device called the "cremailliere", consisting of a piece of notched metal fastened to the stick above the frog, was added as a means of regulating the hair tension.[8] Further improvements for this purpose introduced a screw-button and movable frog by 1790.

The 18th and 19th century musicians, becoming more advanced technically, proved the inadequacy of the primitive versions of bows and demanded a bow which could facilitate their playing, match their instruments, and create different effects to enhance their musicality with strokes such as a smooth legato, wide detache, martellato, staccato, spiccato, sulla tastiera, ponticello, double stops, and chords. Late in the 18th century the Tourtes in Paris and John Kew Dodd in London, each unaware of the other's work, attempted to discover the elements necessary to achieve perfection in the structure of a bow. They both reached the same conclusions regarding the shape, length, hair distribution, weight, and center of gravity. Dodd, however, although the creator of bows noted for their elegance and aesthetic appearance, did not maintain a consistency in the

Fig. 3

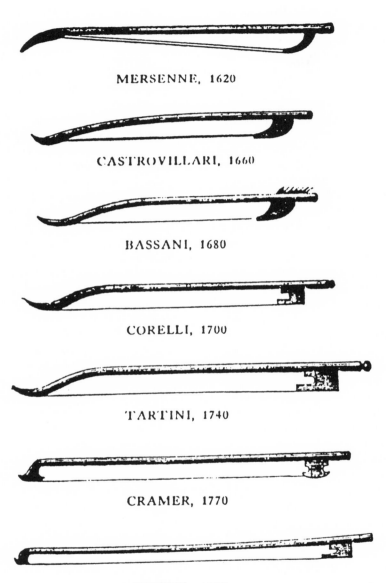

MERSENNE, 1620

CASTROVILLARI, 1660

BASSANI, 1680

CORELLI, 1700

TARTINI, 1740

CRAMER, 1770

VIOTTI, 1790

Display of the Successive Ameliorations of the Bows of the Seventeenth and Eighteenth Centuries, from Joseph Roda's *Bows for Musical Instruments of the Violin Family*, p. 44.

shape of the head, the form and length of the stick, and the mounting of the frog. The Tourte masterpieces serve as a standard of excellence and artistry for the bow just as the Stradivarius remains the zenith of construction for the violin.

In an experiment conducted by Anders Askenfelt, which he reported in an article titled the "Measurement of Bow Motion and Bow Force in Violin Playing,"[9] the variations in bowing parameters were observed through the calculation of bow position, bow velocity, and bow force under normal playing conditions. Askenfelt concluded that the bow force modulated from 0.5 to 1.5 Newtons while the bow velocity was found to be typically between 0.1 and 1.0 meters per second. These two factors, along with the variation of the bow-bridge distance, altered the decibel intensity of the sounds produced. Simply changing the bow-bridge distance may alter the dynamic level up to 18 dB.

When these three variables were held constant (a ± 1.0 N, $\pm .01$ m/s and ± 0.1 cm deviation was occasionally found to exist), a German scientist, Meyer,[10] discovered that the violin has a highly directional sound output, meaning that certain frequencies travel towards specific directions. As a result, while a recording is being made, certain resonances may be overlooked because their radiation pattern leads away from a particular microphone.

The object of this study is — while controlling the velocity, force, and bow-bridge distance manually — to describe and rationalize a three-dimensional polar field of the direction and decibel level of specific frequencies, at the same time introducing additional variables, and noting the resulting discrepancies in the polar patterns, through the use of five different bows and two fundamentally distinctive bow strokes: legato and spiccato.

Endnotes

[1]Joseph Roda, *Bows for Musical Instruments of the Violin Family* (William Lewis and Son, 1959).

[2]David Reck, *Music of the Whole Earth* (Charles Scribner's Sons, 1977).

[3]Emanuel Winternitz, *Musical Instruments and Their Symbolism in Western Art* (Faber and Faber, 1967).

[4]Roda, supra note 1, at 40.

[5]Ibid., at 42.

[6]Winternitz, supra note 3, at 198.

[7]Roda, supra note 1, at 44.

[8]Ibid., at 45.

[9]Anders Askenfelt, *Measurement of Bow Motion and Bow Force in Violin Playing*, Journal of Accoustical Society of America 80(4) October (1986).

[10]Carlee Maley Hutchins, *The Acoustics Of Violin Plates*, (Scientific American, 1981).

•

Peter Steinart

In "Abundance," from Audubon magazine, Peter Steinart takes a look at the human condition via the human dimension.

Abundance

Humankind is probably now the most numerous species of mammal. With five billion of us, no other primate comes close. North American buffalo may once have numbered 60 million and white-tailed deer 40 million. There are probably 10 million wildebeest in Africa. Even livestock fail to outnumber us. There are 1.3 billion cattle, 1.1 million sheep, 787 million pigs, and 64 million horses in the world. And carnivores don't begin to equal us. Though cats and dogs are probably the most numerous, there is only one dog and one cat for every five people in the United States. In other countries the ratio of cats and dogs to people is even smaller.

Among mammals, only rodents are likely to rival humans for the teeming title. It has been said that there is one rat for every person on Earth. But where studies have been made, the rat population comes out smaller. Studies by David Davis in the 1940s found one rat for fifteen people in Baltimore and one for thirty-six in New York City. Since there are 400 species in the *Rattus* genus, and several species (brown and Norway rats for example) are lumped together in these counts, the overall number of any species seems unlikely to reach five billion.

There are other rodent candidates, however. Ernest Thompson Seton estimated that there were once five billion prairie dogs, but today they number a tiny fraction of that. One zoologist believes lemmings, which occupy a vast expanse of Arctic tundra, are more numerous than humans. He has counted them in densities of up to 1,000 per acre and guesses it is safe to say that brown or collared lemmings must number in the trillions.

But there are dissenting voices here. Tony Sinclair, a University of British Columbia zoologist, says, "Trillions is too big a number. I think people

overestimate them. I would doubt that lemmings are even in the billions."
Sinclair believes that lemming populations are divided by mountain ranges
and rivers into noninterbreeding subpopulations. So, on that technicality,
man may be the most numerous species of mammal.

Among birds, only the chicken outnumbers us. With a world population of at least 7.3 billion, chickens appear to be the most numerous land
vertebrate. The passenger pigeon is said to have numbered perhaps three
billion; if that controversial estimate is accurate, it outnumbered humans
while it survived. The most numerous *wild* species today is probably one
of the shearwaters or petrels that swarm in huge flocks over the southern
seas. John Warham of the University of Canterbury in New Zealand estimated the world population of sooty shearwaters at "one thousand million." From there the numbers drop sharply.

U.S. Fish and Wildlife Service studies conclude that North America
produced only 62 million ducks in 1985, a number which takes in ten of
the most noticeable species. Shorebirds, which we see in large flocks, seldom amount to a million in a species. Says Peter Myers of the Philadel-
phia Academy of Natural Sciences, "We see a lot of them in many places,
but because of the habitat we're living in, we're seeing most of them.
They come across as very common, but they're not." House finches,
which we might guess from familiarity are hugely abundant, are village
birds with restricted habitat. No reptiles or amphibians come close to
bird numbers.

Even most insects are less numerous than people. But several species of
ants and bees outnumber us. Says Terry Erwin of the Smithsonian Institution's Department of Entomology, "Some particular species of ant are
estimated to have 10^{18} individuals. In some species of termite on the African savannah, you could take a mound apart and find two million individuals" and go on to find, say thirty such mounds in an acre, and millions
of such acres. "The world record," says Harvard University biologist Edward Wilson, "is now held by a Japanes species, *Formica yessensis*, a single colony of which on the Hokkaido coast had 300 million workers and a
million queens." There are probably a million times more ants than people on Earth. Among aphids and small beetle species which specialize in
ornamental crops and tree species in the northern temperate zone, says
Wilson, it would not be extraordinary to find ten billion individuals.

Erwin believes there are as many as 30 million different species of insect. Says Wilson, "A large number of species are represented at any given
time by a few hundred thousand or a few million. The ants, termites,
social bees, and wasps make up 75 percent of all insect biomass. A small
number are excessively abundant, in the trillions or hundreds of billions.
Most species are relatively uncommon."

The species that are apt to far outnumber us are those we can't easily see. They are subterranean creatures, ocean dwellers, or things too small for us to take much notice of. In the oceans the numbers grow spectacularly large. For decades the North Atlantic fishery alone yielded over 400 million cod a year.

The numbers can be astronomical. Richard Rosenblatt of the Scripps Institution in California observes that at its height the Peruvian anchovy fishery yielded 14 million tons of fish a year to commercial fishermen and another four to six million to seabirds. He does a rough calculation. "Each anchovetta probably weighed a couple of ounces. Say eight fish to a pound. Multiply it out: 320 billion. That's not the whole population, just what's being caught." According to Paul Smith of the National Marine Fisheries Service, just the central subpopulation of northern anchovy, in the coastal waters off California, was more than 25 billion in a recent year.

Anchovies live in coastal waters and may be rare compared with some small, nonschooling fish that live in the open ocean. Marine scientists think a one-inch, silvery, bristle-toothed mid-ocean fish of the genus *Cyclothone* may be the world's most abundant vertebrate, numbering in the trillions. Elbert Ahlstrom of the U.S. Bureau of Commercial Fisheries in 1965 calculated from the number of larvae collected in deep-water hauls that another one-inch mid-ocean fish, *Vinciguerria lucetia*, was even more common.

Smaller sea creatures can be astonishingly abundant. There are probably quadrillions of the Antarctic krill, *Euphausia superba*, a one-inch, shrimp-like creature that is the food of baleen whales and occurs in vast swarms throughout southern waters. Biologists have recorded densities of 35 pounds per cubic yard of ocean. One authority claims that, in the days before whales were hunted, they consumed 150 million tons of krill a year, an annual harvest of something like 45 trillion individuals.

Marine copepods, one-inch shrimp-like crustaceans, are even more abundant. Copepods constitute 70 percent of the plankton in the oceans and are thought to consume half the photosynthetic production of the sea's plants. There are 6,000 known species, but a few of the genus *Calanus* are incredibly abundant and widely distributed. Eggs of one species, *Acartia clausi*, have been counted as thick as 3.4 million per square meter of ocean bottom.

To find such numbers among land creatures, we have to look at smaller and smaller things. David Pimentel of Cornell University estimates there are at least 10,000 insects and billions of mites and collembolans per acre of habitat in upstate New York: "I'd say you could touch a billion mites in a hectare. I could grossly estimate that with a mite species you'd be into the trillions." Bacteria are simply uncountable. "With millions in a handful of soil," says Wilson, "they just dwarf the human population."

Looking at numbers this way is something relatively new to us. Before we wrote books on the subject, we lived more intimately with other creatures and saw them as individuals. Subsistence hunters believe animals have individual spirits which return to Earth in new bodies after a creature is slain for food. So, overall numbers were of little use to early hunters. Even our first formal naturalists, the Greek and Roman writers, seldom considered quantity. Aristotle and Pliny make no mention of animal numbers.

It wasn't until Europeans began exploring distant shores that they discovered abundance. When they reached Africa and the New World, they saw landscapes unaffected by agriculture and firearms, and they encountered life in such numbers that it amazed them. Sir Walter Raleigh declared of the New World that "in all the world the like abundance is not to be found." Coronado, in giving the first European account of the buffalo, remarked not on their size but their abundance: "I found such a quantity, that it is impossible to number them."

In the 17th century the English adventurer Thomas Morton wrote of "millions of turtledoves" in Massachusetts. George and Leonard Calvett wrote of Maryland's "infinite number of birds. . .by all which it appears that the country abounds not only with profit but with pleasure."

An 18th-century visitor to the banks of Newfoundland declared, "It seemed as if all the Fowles of Air were gathered thereunto. They so bemused the eye with their perpetual comings and goings that their number quite defied description. There can be but few places on Earth where is to be seen such a manifestation of the fecundity of His Creation." Peter Martyr wrote in 1516 that explorer John Cabot found off Newfoundland "so great a quantity of great fish that at times they even stayed the passage of his ships."

Counting absolute numbers of such creatures has never appealed to science. Until recently, naturalists were mostly interested in discovering and naming species. It wasn't until the 1870s—when Darwin and others were counting earthworms in soils to show that seemingly insignificant creatures had done more than man to shape the world—that numbers began to play much of a role in biology. Even today, few scientists look at overall numbers. The uncertainties posed by vast territories and differing habitats make accuracy elusive. Says Tony Sinclair, "It's not a very useful question. It's only relevant when you get down to rare animals and probabililties of extinction."

But outside of science, we have tended to see animal numbers as the measure of Earth's abundance. Animal abundance particularly led Americans to believe in material abundance. Martyr declared, "By reason of the rankness and fruitfulness of the ground, kine, swine, and horses do marvelously increase in these regions, and grow to a much bigger quantity." A character in the 1605 play *Eastward Ho!* says of Virginia, "Wild boar there is as common as our tamest bacon, and venison as mutton." The dollar became a "buck" after frontier traders reckoned prices in deerskins, and the buffalo's image appeared on American currency.

The vast numbers of creation in the New World even did much to shape society. They suggested from the start that there were more than enough rewards for those who strove. A democratic system, declared historian David Potter, depends upon the existence of an economic surplus, which in America derived from the natural wealth of the continent. European societies were class-bound especially because they had reached the limits of their natural resources. Observes Potter, "The European mind often assumes implicitly that the volume of wealth is fixed; that most of the potential wealth has already been converted into actual wealth . . . Europe has always conceived of redistribution of wealth as necessitating the expropriation of some and the corresponding aggrandizement of others." But Americans believe wealth still remains to be found, and they address social problems not by taxing the rich but by seeking to increase productive capacity. Because abundance and freedom are so intermingled in the American mind, it is hard to say which one is the means and which the end.

As humankind multiplies itself and reduces the abundance of other creatures available to our eyes, our views of both are likely to change. More and more, the teeming life we see is human, jostling one another on city streets, staring blankly from refugee camps in Africa. Less frequently do we see the sky dark with birds or feel the ground rumble with hoofbeats.

We describe ourselves more in the terms we used to reserve for ants and termites and lemmings. Elliott Norse of the Ecological Society of America says, "I don't think there's anything magical about the number five billion. If people were each the size of a bacterium, we could all conveniently fit on a dead chipmunk and do our thing. Unfortunately, we're not. It's the number of us times our size—in other words, our biomass—that counts."

We see ourselves less and less in the eyes of wolves and bears and eagles. We identify more and more with creatures whose strategies of life entail huge reproductive potential and a carelessness of death. We borrow our ideas of abundance now from a biology that uses numbers to describe a situation that is out of whack, an environment in which human development has squeezed out natural species or overbred insect pests.

It may not be coincidental that, as the buffalo fades from the back of the nickel, we, like the Europeans, now argue who is going to pay the largest share of the tax burden. There is much to be said for having abundance around just to guide our daydreams, to keep us friendly and civil, and to give us hope.

•

Diane Spann

Diane Spann, a junior at the Manhattan Center for Science and Mathematics, wrote her essay over a span of two or three workshops as a series of questions.

Television in Society

I didn't know what to write about when I realized that I would have to pick another topic, then I thought about writing about television and the different types of shows that come on at different times.

Some people watch shows because of their ratings (based on personal behavior or language), some watch soap operas, like housewives, baby-sitters, and people who have sit-down jobs.

Then there are the kids who watch their ABC specials and cartoons. Anyone may watch cartoons. Maybe they watch them because there are kids in the house and they have no choice or maybe there's nothing else on TV that interests them. For instance, there might be a religious program on and they don't want to watch it, so they change the channel.

If so, what type of people watch religious programs? Do they watch ministers on TV because they are too ill to go to church or because they are lazy? Is it because they don't have the kind of clothing they feel they have to wear to church? Maybe they don't want to feel embarrassed because they don't have money to put in the contribution plate. Highly religious people may be the only people who watch religious programs. They may watch them every moment of time they can because their church may be inaccessible at the time. Maybe they only watch those programs to hear the music.

What type of people watch exercise programs? Are they uncontrollably fat? Are they exercise fanatics? Do they watch Channel 7 exercise at 5:20 A.M. or Channel 4 where the girls always jump around in those tight-fitted leotards? Maybe it depends on if they are male or female, or maybe it depends on how hard they want to work out.

116

What type of people stay up till 5:00 A.M. watching TV? What do they watch? Are there people who watch only new programs and people who watch only old reruns?

Why do people become addicted? How do people decide the amount of time they have to watch TV? Do they only watch from 8 P.M. to 11 P.M.? Why do they get home from work late? What do you watch at 8 P.M.? Which is more important to you, "The Cosby Show" or the news? Why do people watch the news? Is there a certain segment of the news that people watch? Do people only watch the sports report, the weather, the local report or national news?

What type of people watch comedy shows? Are they comedians themselves looking for someone to laugh at, or are they humorless people trying to loosen up and gain a sense of humor?

What type of people watch documentaries? Do kids watch them because they have to for school, or do they watch them out of curiosity?

Who watches nature films? Teachers? Students? People who want to learn what makes animals tick? Just nature lovers? Do some people watch nature films for the scenery?

What type of people buy cable TV (pay TV)? Are they people who don't want to pay money to see a movie? Are they disabled and can't make it to the movies? Are they deaf and have to wait until close captioned films come on TV? If so, what do they watch on cable TV? Haven't you ever wondered how TV shows are rated? Of course, it's by the Nielsen ratings, but how does the Nielsen Company place the ratings boxes? Who thought up the system of letting one family provide the rating of a couple thousand?

What about the shows that are rated but aren't given public acknowledgement that they exist? Are there shows such as these? I'm sure a lot of people watch wrestling, to give themselves an outlet for their frustration through the violence of wrestling. Is it because wrestling is funny? Why do people watch boxing? Do they get a thrill out of watching men beat each other up? No? Then what is it? Maybe meek and quiet people watch wrestling and boxing. It could be an outlet for some of the feelings they want to express but can't.

Does wrestling get a Nielsen rating, and why aren't we aware of it? Do you watch shows based on how popular they are, or do you watch shows because you like them? Who decides what shows you watch in your home? What types of people never watch TV? Who are the ones that become addicted?

●

Charles Darwin

Here are two samples from the "Recapitulation and Conclusion" chapter of Charles Darwin's The Origin of Species.

When we no longer look at an organic being as a savage looks at a ship, as something wholly beyond his comprehension; when we regard every production of nature as one which has had a long history; when we contemplate every complex structure and instinct as the summing up of many contrivances, each useful to the possessor, in the same way as any great mechanical invention is the summing up of the labor, the experience, the reason, and even the blunders of numerous workmen; when we thus view each organic being, how far more interesting — I speak from experience — does the study of natural history become!

•

It is interesting to contemplate a tangled bank, clothed with many plants of many kinds, with birds singing on the bushes, with various insects flitting about, and with worms crawling through the damp earth, and to reflect that these elaborately constructed forms, so different from each other, and dependent upon each other in so complex a manner, have all been produced by laws acting around us. These laws, taken in the largest sense, being Growth with Reproduction; Inheritance which is almost implied by reproduction; Variability from the indirect and direct action of the conditions of life, and from use and disuse: a Ratio of Increase so high as to lead to a Struggle for Life, and as a consequence to Natural Selection, entailing Divergence of Character and the Extinction of less-improved forms. Thus, from the war of nature, from famine and death, the most exalted object which we are capable of conceiving, namely, the production of the higher animals, directly follows. There is grandeur in this view of life, with its several powers, having been originally breathed by the Creator into a few forms or into one; and that, whilst this planet has gone cycling on according to the fixed law of gravity, from so simple a beginning endless forms most beautiful and most wonderful have been, and are being evolved.

•

Stephen Jay Gould

Stephen Jay Gould, in his essay "Nonmoral Nature," reminds us that the most pressing problem of natural theology in 1829 was: "If God is benevolent and the Creation displays his 'power, wisdom, and goodness,' then why are we surrounded with pain, suffering, and apparently senseless cruelty in the animal world?" He then illustrates such cruelty with a description of the behavior of the ichneumon fly. At the end of the essay he quotes from a letter Darwin wrote to Asa Gray on the subject. "I feel most deeply that the whole subject is too profound for the human intellect. A dog might as well speculate on the mind of Newton. Let each man hope and believe what he can." Here is a brief selection from Gould's essay.

The ichneumons, like most wasps, generally live freely as adults but pass their larval life as parasites feeding on the bodies of other animals, almost invariably members of their own phylum, Arthropoda. The most common victims are caterpillars (butterfly and moth larvae), but some ichneumons prefer aphids and others attack spiders. Most hosts are parasitized as larvae, but some adults are attacked, and many tiny ichneumons inject their brood directly into the egg of their host.

The free-flying females locate an appropriate host and then convert it to a food factory for their own young. Parasitologists speak of ectoparasitism when the uninvited guest lives on the surface of its host, and endoparasitism when the parasite dwells within. Among endoparasitic ichneumons, adult females pierce the host with the ovipositor and deposit eggs within it. (The ovipositor, a thin tube extending backward from the wasp's rear end, may be many times as long as the body itself.) Usually, the host is not otherwise inconvenienced for the moment, at least until the eggs hatch and the ichneumon larvae begin their grim work of interior excavation. Among ectoparasites, however, many females lay their eggs directly upon the host's body. Since an active host would easily dislodge the egg, the ichneumon mother often simultaneously injects a toxin that paralyzes the caterpillar or other victim. The paralysis may be permanent, and the caterpillar lies, alive but immobile, with the agent of its future destruction secure on its belly. The egg hatches, the helpless caterpillar twitches, the wasp larva pierces and begins its grisly feast.

•

Martin Gardner

In his introduction to an essay by William James, Martin Gardner makes these observations.

To that strange, disquieting question, "Why does anything exist?" science can never hope to provide an answer. The reason is simple. Science can only answer a "why?" by placing an event within the framework of a more general descriptive law. Why does the apple fall? Because of the law of gravitation. Why the law of gravitation? Because of certain equations that are part of the theory of relativity. Should physicists succeed some day in writing one ultimate equation from which all physical laws can be derived, one could still ask, "Why *that* equation?" If physicists reduce all existence to a finite number of particles or waves, one can always ask, "Why *those* particles?" or "Why *those* waves?" There necessarily must remain a basic substratum—a "dark abyss," as Santayana once described it, "before which intelligence must be silent for fear of going mad." It is the Unknowable of Spencer, the Noumena of Kant, the transcendent "Wholly Other" world of Plato and Christianity and all the great religions. It is the Tao that cannot be seen or heard or named because if it could be seen or heard or named it would not be Tao.

Though reason must be silent, the emotions need not be, and it is hard to conceive of a physicist with soul so dead, who never to himself hath said, "This is my own, my native hand!" "The man who cannot wonder," wrote Carlyle, ". . . were he President of innumerable Royal Societies, and carried the whole *Mécanique Céleste* and *Hegel's Philosophy,* and the epitome of all Laboratories and Observatories with their results, in his single head,—is but a Pair of Spectacles behind which there is no Eye. Let those who have Eyes look through him, then he may be useful."

•

Gerald Gourdain

Gerald Gourdain, a sophomore at the Manhattan Center for Science and Math, wrote most of his metaphysical meditation outside of class.

To the Reader
All of the following may or may not hold truth, but before deciding for yourself its truthfulness, base your decision on sound thinking and thoughtful reason.

This scientific essay is based on the mental perception of man and represents his ability to try to understand beyond his understanding. This essay helps one to realize man's unquenched curiosity in the face of the universe, and his never satisfied hunger for knowledge.

This essay is written in a form in which concepts steadily arise, representing man's eagerness to reach higher. The concepts aren't distinctly related, in order to represent man's hunger for knowledge of everything around him. The climax is reached when the concept of God is introduced to represent man's goal.

To understand the essay one must start with an open mind and then judge not during but after reading the essay.

•

Life is a chain in which all that is living and all that exists are related. Everything, whether organic or inorganic, is derived from the same source of matter.

Organisms affect their environment and their environment in turn affects them, both reacting to stimuli, both being stimulators.

A never-ceasing cycle of relations occurs also on a greater level in the universe, even in non-living matter. Planets affect their environments whether they be other planets or the solar winds, which originate from the sun, and vice versa.

The universe is a whole, and anything within it, whether functioning individually or not, is part of the whole and operates as a unit. One might say, "There is a method to the madness," with the word *madness* implying the seeming chaos.

The rocks that orbit the planet Uranus are related to the fine piece of cherry cake sitting in some corner bakery. An obvious idea is that both might have at least one chemical simularity, but my mind perceives an even deeper notion, and that is: if you were able to trace back into time the rocks of Uranus in orbit and the piece of cherry cake sitting, there is no doubt that you would reach a point of relation. It might be the point in which they both were carbon atoms in the middle of space, or even the might-be-fact that they were atoms in the theoretical Big Bang.

The universe is where living matter was born out of non-living matter and where non-living matter is created from living matter. It's a full circle.

Humans themselves are carbon based, and mostly water. We ourselves are a highly organized system of non-living matter, down to the lipids which make up our cell membranes. Our cells are made of chemicals interreacting so that we may function, or, as we say, live. We think and move by electricity, which is dead. Our own definition of non-living, it's our very structure, because there isn't a structure or atom and molecule that we can say isn't somewhere else or part of something that isn't living. Now we can say that the definition of life is highly organized matter working as a unit to preserve and provide for its own needs. Life is the highest form of matter. Life is the highest form of matter that is living.

From the very basic level, we are cells made of chemicals that aren't living, but yet operate together as a unit, are organized, and by our considerations living. The very DNA that makes it up, and the other chemicals, all are non-living.

After all these deductions, we rise to a question: What exactly proves we are alive? We don't really know, or do we? One thing for sure is that we can assume, in scientific terms, some sort of theory or hypothesis. My mind has persuaded me to believe in the spirit, as religions call it; a living part of man, as defined. The spirit is a part that can neither be seen nor touched, a part that is described as the living energy of man, a part that seems to operate on a different plane, an undetectable energy source with its own set of universal rules.

Does this exist? "Yes" seems to be my only moderate answer for now, or even the only answer. If it does exist, does it exist for all organisms? I would say not, because it gives us the highest ability and the most awesome power, the ability to reason.

We are functioning as living structures because we are made up of a whole bunch of non-living structures that prescribe our trait as living. Is not the universe made up of non-living structures, as I said, "operating as a whole"? Consider this: Is the universe as a whole a living organism that we are somehow part of? Well, I believe you can't tell because:

'1) You're (we're) too busy being part of it
and

2) You'd (we'd) have to be on the outside looking in, and, in my own interpretation, that means God.

God represents the power that created the entire Cosmos, again whether it be the Big Bang or some other sort of energy; as I would say, the "most supreme" of all other energy.

God, if He (?) is part of the universe, is just as insignificant as we might be. From my logical considerations, God can't be part of the universe, He has to *be* the universe! Or else he is just another being.

If God is a being, in our eyes he will be high but still insignificant for being part of the universe and not *the* universe. In simple terms, the whole is always greater than the individual or the part, and in this example the whole represents the universe and the individual, or the part, God.

Maybe we can also imply that God is an individual, but an individual that represents the most powerful force in the universe, above all other forces and consistent even if the universe should draw to a close.

Since we have assumed that God exists, then He must be the most unscientific and impossible form of existence, defying all existing science but yet the only true science.

If God is a being, then He is insignificant, but if He is the universe then He is the "Method behind the madness." God, if He exists, is power, power to defy any and all laws of science and reason. God, if He exists, can't be defined as living or non-living.

The only reason we deny such possibilities, in my opinion, is because we believe that reason is the truth and the answer to all, and that everything is logical. Well, if that is true, imagine how easy life would be. Science is a logical learning approach to the universe but is it the ONLY ONE?

•

Joseph D. Ciparick

Joseph D. Ciparick, a high school teacher of chemistry and physics, penned this meditation, which discovers the gods of chemistry and then muses on the mysterious nature of theories.

Chemists on Mt. Olympus
When the Greeks saw magic, they said it was the work of the gods. If chemistry is magical, who is the Greek god of chemistry?

Millenia ago someone somewhere saw grey molten iron appearing in the fire, coming from the red rocks. Magic! A transformation such as this could not be done by mere man. The Greeks attributed it to the god

Hephaestus (also called Vulcan), the god of the fiery volcanos and one of the Greek gods of chemistry.

What? You say there is *no* Greek god of chemistry? I say there are three: the residents of Mount Olympus who preside over magical transformations.

The second god of chemistry — or rather, goddess — is Ceres, the goddess of grain. Why grain and not the other abundant fruits of the earth? Because the other fruits can be plucked and used as they are. But we don't eat grain. Grain is ground into flour and mixed with yeast and water and is transformed by fire, most mysteriously, into bread. To bake a loaf of bread is to make magic.

I would place both Hephaestus and Ceres on the highest magical level. Who, looking at a red lump of rock, would ever envisage getting iron metal? Who, looking at a basket of wheat, could envisage a loaf of bread? A master chemist?

Notice that despite the wealth and glitter of gold, there is no goldsmith on Mount Olympus. Gold is plucked from the earth as it is. Humble iron is more magical.

The third representative of the science of chemistry on Mount Olympus is Dionysus, the god of wine. His claim to fame is not as spectacular as those of Hephaestus and Ceres, since I'm sure that fermentation was discovered by accident when someone tasted a wrinkled grape on a vine long after the harvest was over.

Rocks to iron, wheat to bread, grapes to wine. Today, of course, we know that these transformations are not magic but can be explained by the science of chemistry. We know that reduction of a metal from its oxide involves an exchange of electrons. Or do we? *Scientia* is Latin for "knowledge," but we often confuse what we know with what we believe. We *know* through observation that iron can be gotten from iron ore. We *believe* in the theory that tries to explain why this happens.

What we see is real, but mysterious. Theories are attempts to explain what we see, but they are constructions of man and do not exist outside of his mind and his textbooks. But, you say, even if theories are not real, at least they are *true*. False. It is tempting to conclude that, like observations which can be either accurate or in error, theories can be true or false. Never! Theories are not true or false, any more than Hamlet or a Picasso painting is true or false. Theories are human insights, creative, beautiful, or — when they miss the mark — ugly. But we can never say they are true or false because they are a part of man, not of nature.

Why should molten iron come from a red rock? The fact is that, under certain circumstances, it does. *Why* it does is a mystery. You may talk of iron ore and reducing agents and equilibrium temperatures but, at the core, this reaction is no less mysterious today than it was centuries ago. Unfortunately, we think we understand why reactions take place. And because we believe the theories are true, the magic has gone from our everyday experiences.

Bread and wine and iron and fire and water are still mysteries. Stand humbly before each, and never think that you can evict it from the realm of the mysterious by categorizing it as a redox reaction or enzyme process. Don't replace worship of Greek gods with worship of theories. And savor the richness of our theories. Theories add beauty to nature's mysteries. But no theory ever made a mystery vanish.

•

Carl Sagan

Here are some excerpts from Carl Sagan's "Can We Know the Universe? Reflections on a Grain of Salt."

Science is a way of thinking much more than it is a body of knowledge. Its goal is to find out how the world works, to seek what regularities there may be, to penetrate to the connections of things. . . .

•

If you spend any time spinning hypotheses, checking to see whether they make sense, whether they conform to what else we know, thinking of tests you can pose to substantiate or deflate your hypotheses, you will find yourself doing science. And as you come to practice this habit of thought more and more you will get better and better at it. To penetrate into the heart of the thing—even a little thing, a blade of grass, as Walt Whitman said—is to experience a kind of exhilaration that, it may be, only human beings of all the beings on this planet can feel. We are an intelligent species and the use of our intelligence quite properly gives us pleasure. In this respect the brain is like a muscle. When we think well, we feel good. Understanding is a kind of ecstasy. . . .

•

Fortunately for us, we live in a universe that has at least important parts that are knowable. Our commonsense experience and our evolutionary history have prepared us to understand something of the workaday world. When we go into other realms, however, common sense and ordinary intuition turn out to be highly unreliable guides. It is stunning that as we go close to the speed of light our mass increases indefinitely, we shrink toward zero thickness in the direction of motion, and time for us comes as near to stopping as we would like. Many people think that this is silly, and every week or two I get a letter from someone who complains to me about it. But it is a virtually certain consequence not just of experiment but also of Albert Einstein's brilliant analysis of space and time called the Special Theory of Relativity. It does not matter that these effects seem unreasonable to us. We are not in the habit of traveling close to the speed of light. The testimony of our common sense is suspect at high velocities.

•

José Ortega y Gasset

The Spanish philosopher José Ortega y Gasset (1883-1955), in his essay ''The Barbarism of 'Specialization,''' makes observations that could easily be applied to America in the 1980s. It is a good argument for writing-across-the-curriculum.

The investigator who has discovered a new fact of Nature must necessarily experience a feeling of power and self-assurance. With a certain apparent justice he will look upon himself as "a man who knows." And in fact there is in him a portion of something which, added to many other portions not existing in him, does really constitute knowledge. This is the true inner nature of the specialist, who in the first years of this century has reached the wildest stage of exaggeration. The specialist "knows" very well his own, tiny corner of the universe; he is radically ignorant of all the rest.

Here we have a precise example of this strange new man, whom I have attempted to define, from both of his two opposite aspects. I have said that he was a human product unparalleled in history. The specialist serves as a striking concrete example of the species, making clear to us the radical nature of the novelty. For, previously, men could be divided simply into the learned and the ignorant, those more or less the one, and those more or less the other. But your specialist cannot be brought in under either of these two categories. He is not learned, for he is formally ignorant of all that does not enter into his specialty; but neither is he ignorant, because he is "a scientist," and "knows" very well his own tiny portion of the universe. We shall have to say that he is a learned igno-ramus, which is a very serious matter, as it implies that he is a person who is ignorant, not in the fashion of the ignorant man, but with all the petu-lance of one who is learned in his own special line.

And such in fact is the behavior of the specialist. In politics, in art, in social usages, in the other sciences, he will adopt the attitudes of primitive, ignorant man; but he will adopt them forcefully and with self-sufficiency, and will not admit of—this is the paradox—specialists in those matters. By specializing him, civilization has made him hermetic and self-satisfied within his limitations; but this very inner feeling of dominance and worth will induce him to wish to predominate outside his specialty. The result is that even in this case, representing a maximum of qualification in man—specialization—and therefore the thing most opposed to the mass-man, the result is that he will behave in almost all spheres of life as does the un-qualified, the mass-man.

This is no mere wild statement. Anyone who wishes can observe the stupidity of thought, judgment, and action shown today in politics, art, religion, and the general problems of life and the world by the "men of science," and of course, behind them, the doctors, engineers, financiers, teachers, and so on. That state of "not listening," of not submitting to higher courts of appeal which I have repeatedly put forward as charac-teristic of the mass-man, reaches its height precisely in these partially qualified men. They symbolize, and to a great extent constitute, the ac-tual dominion of the masses, and their barbarism is the most immediate cause of European demoralization. Furthermore, they afford the clear-est, most striking example of how the civilization of the last century, *abandoned to its own devices*, has brought about this rebirth of primitiv-ism and barbarism.

The most immediate result of this *unbalanced* specialization has been that today, when there are more "scientists" than ever, there are far fewer "cultured" men than, for example, about 1750. And the worst is that with these turnspits of science not even the real progress of science itself is as-sured. For science needs from time to time, as a necessary regulator of its own advance, a labor of reconstitution, and, as I have said, this demands an effort towards unification, which grows more and more difficult, in-volving, as it does, ever-vaster regions of the world of knowledge. Newton was able to found his system of physics without knowing much philosophy, but Einstein needed to saturate himself with Kant and Mach before he could reach his own keen synthesis. Kant and Mach—the names are mere symbols of the enormous mass of philosophic and psychological thought which has influenced Einstein—have served to *liberate* the mind of the latter and leave the way open for his innovation. But Einstein is not sufficient. Physics is entering on the gravest crisis in its history, and can only be saved by a new "encyclopaedia" more systematic than the first.

The specialization, then, that has made possible the progress of experimental science during a century, is approaching a stage where it can no longer continue its advance unless a new generation undertakes to provide it with a more powerful form of turnspit.

But if the specialist is ignorant of the inner philosophy of the science he cultivates, he is much more radically ignorant of the historical conditions requisite for its continuation; that is to say: how society and the heart of man are to be organized in order that there may continue to be investigators. The decrease in scientific vocations noted in recent years, to which I have alluded, is an anxious symptom for anyone who has a clear idea of what civilization is, an idea generally lacking to the typical "scientist," the highpoint of our present civilization. He also believes that civilization *is there* in just the same way as the earth's crust and the forest primeval.

•

Sharon Begley

In the "Science" column of Newsweek, journalist Sharon Begley notes a bias in science that affects who is employed in the laboratory, what is chosen for research, and how it is researched.

Liberation in the Lab: A Distaff Science?

Time was when women were as scarce in the laboratory as rats who die of old age. The bias against women in science came under attack in the 1970s, and the number of women earning doctorates in science and engineering began to rise. Now some of these female scientists are examining what they suspect is the result of the years of men-only research. Male domination of science, they say, has affected its very content — what questions are asked and answers found. "Our concern is, what is the effect of the doer of science on the science that gets done?" says Shirley Malcolm of the American Association for the Advancement of Science.

The idea that scientific theories are colored by cultural values is hardly new. In his seminal 1962 book, *The Structure of Scientific Revolutions*, Thomas Kuhn argued that the prevailing "paradigm" in a science depends not only on experimental "truths" but also on the scientific community's shared beliefs. Now scholars are asking if gender shapes those beliefs. In her recent book, *Reflections on Gender and Science*, Evelyn Fox Keller of Northeastern University suggests that values our culture associates with men helped bring modern science into being: the 17th-century figure Francis Bacon, for example, brought to science such masculine values as control and domination—values that still shape scientific thought today. "Our 'laws of nature' are more than simple expressions of the results of objective inquiry," writes Keller. "They must also be read for their personal — and by tradition, masculine — content."

The study of primates has been particularly tinged by masculine attitudes. Primatologists returned from the field with reports highlighting the role of the dominant male—how he fought, mated and, most important, decided who would reproduce. Women venturing into the field bring back a different story. Sarah Hrdy of the University of California finds that among langurs females choose their own mates in order to "maximize chances" that their offspring will survive. Women primatologists have emphasized behaviors other than those involving the male-female hierarchies, mother-and-child interactions—and drew a richer picture of primate society. "Maybe these other things were ignored because of the expectations [male primatologists] brought to their study," says Malcolm.

The themes of domination and control can be found in more arcane research, too. In molecular biology, for example, DNA was held to control the cell while being itself free from outside influence. But beginning in the 1950s, Barbara McClintock of the Cold Spring Harbor Lab challenged this idea. Her experiments with maize showed that gene fragments jump among chromosomes in a spontaneous but apparently coordinated way. McClintock thought these movements were subject more to the control of the organism as a whole or to environmental stress—drought, for instance —than to the DNA. By showing that DNA was itself subject to reprogramming, she implied that DNA was far from being the grandmaster of cellular discipline. Says Keller, "The capacity of organisms to reprogram their own DNA . . . confirms the existence of forms of order more complex than we have . . . been able to account for." McClintock's findings had no place in a paradigm built around "master" DNA; eventually, however, she won the 1983 Nobel Prize in Medicine.

Expecting organisms to be controlled by master molecules does not necessarily produce wrong answers, but incomplete ones (just as theories colored by "female" attitudes might). Take the slime mold. It exists as single cells when food is plentiful, but if its sustenance grows scarce, the cells clump together, differentiate, and form food-hunting aggregates something like amoebas. How do the cells know to get together for the search? Keller and Lee Segel developed a mathematical model of how the

129

cells could organize themselves spontaneously, but it was swamped by a rival theory: that a master cell sent out signals telling the others what to do. While there is no evidence for such a "pacemaker," the paradigm of a master molecule calls for one—leaving unexplained *how* the slime mold differentiates in response to its commands. The pacemaker, says Keller, "provides an unusually simple instance of a predisposition to . . . posit a single central governor."

Attitudes that may lead to "male" and "female" science show up early. In a study of 200 children in computer classes, Sherry Turkle of MIT identified two programming styles. One is "hard mastery," the mastery of the engineer, marked by preplanning and the desire to impose one's will over the machine. The other is "soft mastery," the mastery of the artist, marked by interaction. "A computer program reflects the programmer's mind," says Turkle. By age 10, she says, girls in our culture prefer the interactive approach (and are criticized for programming the "wrong" way) while boys like to plan programs every step of the way.

The gender question has not attracted much notice from scientists: laboratory-bound, they tend not to be interested in the philosophy of their subject. But students of science as a mode of thought have welcomed the feminists' critiques. Ian Hacking of the University of Toronto calls the research a way "to start all of us thinking in new directions" — freeing science to explore avenues of research that the sons of Francis Bacon have ignored.

•

Edgar Allan Poe

Science assumes many identities. As might be expected from the master of horror and suspense, Edgar Allan Poe (1809-49) sees it as a vulture and a thief.

To Science
Science! true daughter of Old Time thou art!
    Who alterest all things with thy peering eyes.
Why preyest thou thus upon the poet's heart,
    Vulture, whose wings are dull realities?
How should he love thee? or how deem thee wise,
    Who wouldst not leave him in his wandering
To seek for treasure in the jewelled skies,
    Albeit he soared with an undaunted wing?

Hast thou not dragged Diana from her car?
And driven the Hamadryad from the wood
To seek a shelter in some happier star?
Hast thou not torn the Naiad from her flood,
The Elfin from the green grass, and from me
The summer dream beneath the tamarind tree?

•

Dana Levine

A senior at Townsend Harris High School at Queens College, Dana Levine gave primary students two stories—one highly alliterative, the other phonetically regular—and tested the effect. This is an excerpt from her paper, which was a winner of the 1988 Westinghouse Science Talent Search contest.

The initial data was collected in two stages, spaced two days apart. During the first session, the children were asked to listen to two stories, one composed of alliteration and one of a normal phonetic pattern. After hearing these twice, the children were given questionnaires, on which they were asked to answer two inferential concept questions and one basic context memory question for each story. They also chose which story they felt was longer and rated the stories on the basis of how much they liked each. Enjoyment was measured by the Smiley Scale.

During the next session, two days later, the children heard short summaries of both stories. They did not hear either story in its original form. They were asked to answer four context questions and one concept question on the alliterated story and four context questions and two concept questions on the non-alliterated story.

All questions were multiple choice, with five answer options. I questioned only one vital, the age of the child, but asked also for the name of the child for organizational purposes. I designed the questionnaire and found it was not ideal. The concept and context questions, though on the same level, were not identical in Sessions One and Two. This prohibited testing absolute memory or conceptual knowledge by individual variable. Instead, raw total scores of correct answers by type were analyzed. In order to obtain totals, the value of the concept question in Session Two for the alliterated story had to be weighted in order to help it correspond to the two concept questions that were answered in every other section.

One school was chosen for a further study, conducted two and a half months after the initial session. During this session, the children again heard the summaries of both stories. They were then asked to answer four context questions. These questions were identical to those they had seen in the two previous sessions. This extra session was used to test long-term memory of both stories.

The Alliterated Story

Wednesday, Wendy watched William walk with a walrus. The walrus waddled and wiggled when he walked. The walrus was wearing a white wristwatch. Walking next to William was Walter, the watchdog.

While they walked down Willow Way, they wondered why the weather was so warm that Wednesday. They all wanted water. The walrus wished for water to swim in. Wendy and William wanted water to drink. Walter the watchdog wanted water to wash with.

"I wish it were winter!" William whined.

"Where can we find water?" Wendy wondered.

"Well . . . ," the walrus said, "I'll waive my magic wand and we'll get water."

"No way!" Walter whispered.

"Don't worry. With my magic wand, the water we want will be ours."

He waived his wand and without warning, a wagon full of water wheeled where they were.

"Wow! Wonderful wet water!" said Wendy.

"Thanks, Walrus."

"You're welcome," said Walrus.

The Phonetically Regular Story

One day Danny was playing with a red beach ball. It was a birthday present from his mother. The beach ball was very big and shiny. He wanted to play with it all alone, without sharing with anyone else.

Lisa and Jennifer saw Danny playing with his beautiful, big, red beach ball. They wanted to play too. Danny was having so much fun with his new toy. They asked him nicely if they could play with it.

"I don't know. It is new."

"We know, so we'll be careful," Lisa said.

"I don't think so," Danny told them.

"Please let us, Danny. It looks like so much fun," Jennifer asked him.

"I want to play," Danny said.

"We could all play catch together," Lisa and Jennifer told him.

Danny liked this idea very much and they formed a circle for three-way catch.

"This is much better than playing alone."

"Yes, Danny."

"We knew you would think so."

•

John Burroughs

Here is naturalist John Burroughs (1837-1921) on scientist Darwin. This piece of writing thrusts us into the middle of that lively debate of our times—Evolution vs. Creationism—which was a debate in his as well. The thoughts here are the inspiration of some and the dread of others in that debate.

It is too true that the ornithologists of our day for the most part look upon the birds only as so much legitimate game for expert dissection and classification, and hence have added no new lineaments to Audubon's and Wilson's portraits. Such a man as Darwin was full of what we may call the sentiment of science. Darwin was always pursuing an idea, always tracking a living, active principle. He is full of the ideal interpretation of fact, science fired with faith and enthusiasm, the fascination of the power and mystery of nature. All his works have a human and almost poetic side. They are undoubtedly the best feeders of literature we have yet had from the field of science. His book on the earthworm, or on the formation of vegetable mold, reads like a fable in which some high and beautiful philosophy is clothed. How alive he makes the plants and the trees! — shows all their movements, their sleeping and waking, and almost their very dreams — does, indeed, disclose and establish a kind of rudimentary soul or intelligence in the tip of the radicle of plants. No poet has ever made the trees so human. Mark, for instance, his discovery of the value of cross-fertilization in the vegetable kingdom, and the means Nature takes to bring it about. Cross-fertilization is just as important in the intellectual kingdom as in the vegetable. The thoughts of the recluse finally become pale and feeble. Without pollen from other minds, how can one have a race of vigorous seedlings of his own? Thus all Darwinian books have to me a literary or poetic substratum. The old fable of metamorphosis and transformation he illustrates afresh in his *Origin of Species*, in the *Descent of Man*. Darwin's interest in nature is strongly scientific, but our interest in him is largely literary; he is tracking a principle, the principle of organic life, following it through all its windings and

133

turnings and doublings and redoublings upon itself, in the air, in the earth, in the water, in the vegetable, and in all the branches of the animal world; the footsteps of creative energy; not why, but how; and we follow him as we would follow a great explorer, or general, or voyager like Columbus, charmed by his candor, dilated by his mastery. He is said to have lost his taste for poetry, and to have cared little for what is called religion. His sympathies were so large and comprehensive; the mere science in him is so perpetually overarched by that which is not science, but faith, insight, imagination, prophecy, inspiration, — "substance of things hoped for, the evidence of things not seen;" his love of truth so deep and abiding; and his determination to see things, facts, in their relations, and as they issue in principle, so unsleeping, — that both his poetic and religious emotions, as well as his scientific proclivities, found full scope, and this demonstration becomes almost a song. It is easy to see how such a mind as Goethe's would have followed him and supplemented him, not from its wealth of scientific lore, but from its poetic insight into the methods of nature.

•

William Wordsworth

This passage from William Wordsworth's The Excursion *is an echo, or, perhaps more precisely, the sound that became the echo, of John Burroughs' idea that "until science is mixed with emotion, and appeals to the heart and imagination, it is like dead inorganic matter; and when it becomes so mixed and so transformed it is literature."*

And further; by contemplating these Forms
In the relations which they bear to man,
He shall discern, how, through the various means
Which silently they yield, are multiplied
The spiritual presences of absent things.
Trust me, that for the instructed, time will come
When they shall meet no object but may teach
Some acceptable lesson to their minds
Of human suffering, or of human joy.
So shall they learn, while all things speak of man,
Their duties from all forms; and general laws,

And local accidents, shall tend alike
To rouse, to urge; and, with the will, confer
The ability to spread the blessings wide
Of true philanthropy. The light of love
Not failing, perseverance from their steps
Departing not, for them shall be confirmed
The glorious habit by which sense is made
Subservient still to moral purposes,
Auxiliar to divine. That change shall clothe
The naked spirit, ceasing to deplore
The burthen of existence. Science then
Shall be a precious visitant; and then,
And only then, be worthy of her name;
For then her heart shall kindle; her dull eye,
Dull and inanimate, no more shall hang
Chained to its object in brute slavery;
But taught with patient interest to watch
The processes of things, and serve the cause
Of order and distinctness, not for this
Shall it forget that its most noble use,
Its most illustrious province, must be found
In furnishing clear guidance, a support
Not treacherous, to the mind's *excursive* power.
—So build we up the Being that we are;
Thus deeply drinking-in the soul of things
We shall be wise perforce; and, while inspired
By choice, and conscious that the Will is free,
Shall move unswerving, even as if impelled
By strict necessity, along the path
Of order and of good. Whate'er we see,
Or feel, shall tend to quicken and refine;
Shall fix, in calmer seats of moral strength,
Earthly desires; and raise, to loftier heights
Of divine love, our intellectual soul.

•

'To every Form of being is assigned,'
Thus calmly spake the venerable Sage,
'An *active* Principle:—howe'er removed
From sense and observation, it subsists

In all things, in all natures; in the stars
Of azure heaven, the unenduring clouds,
In flower and tree, in every pebbly stone
That paves the brooks, the stationary rocks,
The moving waters, and the invisible air.
Whate'er exists hath properties that spread
Beyond itself, communicating good,
A simple blessing, or with evil mixed;
Spirit that knows no insulated spot,
No chasm, no solitude; from link to link
It circulates, the Soul of all the worlds.
This is the freedom of the universe;
Unfolded still the more, more visible,
The more we know; and yet is reverenced least,
And least respected in the human Mind,
Its most apparent home. . .'

•

J. Robert Oppenheimer

J. Robert Oppenheimer (1904-67), administrator, mathematician, physicist, student of Greek and Sanskrit, and director of the Los Alamos laboratories at the time of the development of the atom bomb, shares his uniquely informed thoughts on the responsibilities of scientists in these excerpts from his lecture, "Physics in the Contemporary World."

No scientist can hope to evaluate what his studies, his researches, his experiments may in the end produce for his fellow men, except in one respect —if they are sound, they will produce knowledge. And this deep complementarity between what may be conceived to be the social justification of science and what is for the individual his compelling motive in its pursuit makes us look for other answers to the question of the relation of science to society.

One of these is that the scientist should assume responsibility for the fruits of his work. I would not argue against this, but it must be clear to all of us how very modest such assumption of responsibility can be, how very ineffective it has been in the past, how necessarily ineffective it will surely be in the future. In fact, it appears little more than exhortation to the man of learning to be properly uncomfortable, and, in the worst instances, is used as a sort of screen to justify the most casual, unscholarly, and, in the last analysis, corrupt intrusion of scientists into other realms of which they have neither experience nor knowledge, nor the patience to obtain them.

The true responsibility of a scientist, as we all know, is to the integrity and vigor of his science. And because most scientists, like all men of learning, tend in part also to be teachers, they have a responsibility for the communication of the truths they have found. This is at least a collective if not an individual responsibility. That we should see in this any insurance that the fruits of science will be used for man's benefit, or denied to man when they make for his distress or destruction, would be a tragic naiveté.

•

And here, from the same essay by Oppenheimer, are some thoughts on how science should be taught.

. . . There is all the difference in the world between hearing about science or its results and sharing in the experience of the scientist himself and of that of the scientific community. We all know that an awareness of this, and an awareness of the value of science as method, rather than science as doctrine, underlies the practices of teaching to scientist and layman alike. For surely the whole notion of incorporating a laboratory in a high school or college is a deference to the belief that not only what the scientist finds but how he finds it is worth learning and teaching and worth living through.

Yet there is something fake about all this. No one who has had to do with elementary instruction can have escaped a sense of artificiality in the way in which students are led, by the calculations of their instructors, to follow paths which will tell them something about the physical world. Precisely that groping for what is the appropriate experiment, what are the appropriate terms in which to view subtle or complex phenomena, which are the substance of scientific effort, almost inevitably are distilled out of it by the natural patterns of pedagogy. The teaching of science to laymen is not wholly a loss; and here perhaps physics is an atypically bad example. But surely they are rare men who, entering upon a life in which science plays no direct part, remember from their early courses in physics what science is like or what it is good for. The teaching of science is at its best when it is most like an apprenticeship.

[Harvard] President Conant, in his sensitive and thoughtful book *On Understanding Science*, has spoken at length of these matters. He is aware of how false it is to separate scientific theory from the groping, fumbling, tentative efforts which lead to it. He is aware that it is science as method and not as doctrine which we should try to teach. His basic suggestion is that we attempt to find, in the history of our sciences, stories which can be re-created in the instruction and experiment of the student and which thus can enable him to see at firsthand how error may give way to less error, confusion to less confusion, and bewilderment to insight.

•

Jean Henri Fabré

Jean Henri Fabré (1823-1915), in his essay "The Sacred Beetle," finds dung to be as high-minded a scientific subject as any, and shows that science may be impartial without being cold in its enthusiasms.

Who is this that comes trotting towards the heap, fearing lest he reach it too late? His long legs move with awkward jerks, as though driven by some mechanism within his belly; his little red antennae unfurl their fan, a sign of anxious greed. He is coming, he has come, not without sending a few banqueters sprawling. It is the Sacred Beetle, clad all in black, the biggest and most famous of our Dung-beetles. Behold him at table, beside his fellow-guests, each of whom is giving the last touches to his ball with the flat of his broad fore-legs or else enriching it with yet one more layer before retiring to enjoy the fruit of his labors in peace. Let us follow the construction of the famous ball in all its phases.

The clypeus, or shield, that is the edge of the broad, flat head, is notched with six angular teeth arranged in a semi-circle. This constitutes the tool for digging and cutting up, the rake that lifts and casts aside the unnutritious vegetable fibers, goes for something better, scrapes and collects it. A choice is thus made, for these connoisseurs differentiate between one thing and another, making a rough selection when the Beetle is occupied with his own provender, but an extremely scrupulous one when it is a matter of constructing the maternal ball, which has a central cavity in which the egg will hatch. Then every scrap of fiber is conscientiously rejected and only the stercoral quintessence is gathered as the material for building the inner layer of the cell. The young larva, on issuing from the egg, thus finds in the very walls of its lodging a food of special delicacy which strengthens its digestion and enables it afterwards to attack the coarse outer layers.

Where his own needs are concerned, the Beetle is less particular and contents himself with a very general sorting. The notched shield then does its scooping and digging, its casting aside and scraping together more or less at random. The fore-legs play a mighty part in the work. They are flat, bow-shaped, supplied with powerful nervures and armed on the outside with five strong teeth. If a vigorous effort be needed to remove an obstacle or to force a way through the thickest part of the heap, the Dung-beetle makes use of his elbows, that is to say, he flings his toothed legs to right and left and clears a semicircular space with an energetic sweep. Room once made, a different kind of work is found for these same limbs: they collect armfuls of the stuff raked together by the shield and push it under the insect's belly, between the four hinder legs. These are formed for the turner's trade. They are long and slender, especially the last pair, slightly bowed and finished with a very sharp claw. They are at once recognized as compasses, capable of embracing a globular body in their curved branches and of verifying and correcting its shape. Their function is, in fact, to fashion the ball.

Armful by armful, the material is heaped up under the belly, between the four legs, which, by a slight pressure, impart their own curve to it and give it a preliminary outline. Then, every now and again, the rough-hewn pill is set spinning between the four branches of the double pair of spherical compasses; it turns under the Dung-beetle's belly until it is rolled into a perfect ball. Should the surface layer lack plasticity and threaten to peel off, should some too-stringy part refuse to yield to the action of the lathe, the fore-legs touch up the faulty places; their broad paddles pat the ball to give consistency to the new layer and to work the recalcitrant bits into the mass.

Under a hot sun, when time presses, one stands amazed at the turner's feverish activity. And so the work proceeds apace: what a moment ago was a tiny pellet is now a ball the size of a walnut; soon it will be the size of an apple. I have seen some gluttons manufacture a ball the size of a man's fist. This indeed means food in the larder for days to come!

·

Alfred North Whitehead

"A clash of doctrines is not a disaster—it is an opportunity," said philosopher *Alfred North Whitehead (1861-1947) in his* Science and the Modern World. *He applies the idea to the clash between religion and science, drawing an example from scientific developments.*

A clash of doctrines is not a disaster—it is an opportunity. I will explain my meaning by some illustrations from science. The weight of an atom of nitrogen was well known. Also it was an established scientific doctrine that the average weight of such atoms in any considerable mass will be always the same. Two experimenters, the late Lord Rayleigh and the late Sir William Ramsay, found that if they obtained nitrogen by two different methods, each equally effective for that purpose, they always observed a persistent slight difference between the average weights of the atoms in the two cases. Now I ask you, would it have been rational of these men to have despaired because of this conflict between chemical theory and scientific observation? Suppose that for some reason the chemical doctrine had been highly prized throughout some district as the foundation of its social order;—would it have been wise, would it have been candid, would it have been moral, to forbid the disclosure of the fact that the experiments produced discordant results? Or, on the other hand, should Sir William Ramsay and Lord Rayleigh have proclaimed that chemical theory was now a detected delusion? We see at once that either of these ways would have been a method of facing the issue in an entirely wrong spirit. What Rayleigh and Ramsay did was this: they at once perceived that they had hit upon a line of investigation which would disclose some subtlety of chemical theory that had hitherto eluded observation. The discrepancy was not a disaster: it was an opportunity to increase the sweep of chemical knowledge. You all know the end of the story: finally argon was discovered, a new chemical element which had lurked undetected, mixed with the nitrogen. But the story has a sequel which forms my second illustration. This discovery drew attention to the importance of observing accurately minute differences in chemical substances as obtained by different methods. Further researches of the most careful accuracy were undertaken. Finally another physicist, F.W. Aston, working in the Cavendish Laboratory at Cambridge in England, discovered that even the same element might assume two or more distinct forms, termed *isotopes*, and that the law of the constancy of average atomic weight holds for each of these forms, but as between the different isotopes differs slightly. The research has effected a great stride in the power of chemical theory, far transcending in importance the discovery of argon from which it originated. The moral of these stories lies on the surface, and I will leave to you their application to the case of religion and science.

•

Rachel Carson

Rachel Louise Carson (1907-1964), in her exemplary book, The Sea Around Us, *mixes questions and speculations poetically and scientifically.*

In 1946, however, the United States Navy issued a significant bulletin. It was reported that several scientists, working with sonic equipment in deep water off the California coast, had discovered a widespread 'layer' of some sort, which gave back an answering echo to the sound waves. This reflecting layer, seemingly suspended between the surface and the floor of the Pacific, was found over an area 300 miles wide. It lay from 1000 to 1500 feet below the surface. The discovery was made by three scientists, C. F. Eyring, R. J. Christensen, and R. W. Raitt, aboard the U.S.S. *Jasper* in 1942, and for a time this mysterious phenomenon, of wholly unknown nature, was called the ECR layer. Then in 1945 Martin W. Johnson, marine biologist of the Scripps Institution of Oceanography, made a further discovery which gave the first clue to the nature of the layer. Working aboard the vessel, *E. W. Scripps*, Johnson found that whatever sent back the echoes moved upward and downward in rhythmic fashion, being found near the surface at night, in deep water during the day. This discovery disposed of speculations that the reflections come from something inanimate, perhaps a mere physical discontinuity in the water, and showed that the layer is composed of living creatures capable of controlled movement.

From this time on, discoveries about the sea's 'phantom bottom' came rapidly. With widespread use of echo-sounding instruments, it has become clear that the phenomenon is not something peculiar to the coast of California alone. It occurs almost universally in deep ocean basins—drifting by day at a depth of several hundred fathoms, at night rising to the surface, and again, before sunrise, sinking into the depths.

On the passage of the U.S.S. *Henderson* from San Diego to the Antarctic in 1947, the reflecting layer was detected during the greater part of each day, at depths varying from 150 to 450 fathoms, and on a later run from San Diego to Yokosuka, Japan, the *Henderson's* fathometer again recorded the layer every day, suggesting that it exists almost continuously across the Pacific.

During July and August 1947, the U.S.S. *Nereus* made a continuous fathogram from Pearl Harbor to the Arctic and found the scattering layer over all deep waters along this course. It did not develop, however, in the shallow Bering and Chuckchee seas. Sometimes in the morning, the *Nereus'* fathogram showed two layers, responding in different ways to the growing illumination of the water; both descended into deep water, but there was an interval of twenty miles between the two descents.

Despite attempts to sample it or photograph it, no one is sure what the layer is, although the discovery may be made any day. There are three principal theories, each of which has its group of supporters. According to these theories, the sea's phantom bottom may consist of small planktonic shrimps, of fishes, or of squids.

As for the plankton theory, one of the most convincing arguments is the well-known fact that many plankton creatures make regular vertical migrations of hundreds of feet, rising toward the surface at night, sinking down below the zone of light penetration very early in the morning. This is, of course, exactly the behavior of the scattering layer. Whatever composes it is apparently strongly repelled by sunlight. The creatures of the layer seem almost to be held prisoner at the end—or beyond the end—of the sun's rays throughout the hours of daylight, waiting only for the welcome return of darkness to hurry upward into the surface waters. But what is the power that repels; and what the attraction that draws them surfaceward once the inhibiting force is removed? Is it comparative safety from enemies that makes them seek darkness? Is it more abundant food near the surface that lures them back under cover of night?

Those who say that fish are the reflectors of the sound waves usually account for the vertical migrations of the layer as suggesting that the fish are feeding on planktonic shrimp and are following their food. They believe that the air bladder of a fish is, of all structures concerned, most likely from its construction to return a strong echo. There is one outstanding difficulty in the way of accepting this theory: we have no other evidence that concentrations of fish are universally present in the oceans. In fact, almost everything else we know suggests that the really dense populations of fish live over the continental shelves or in certain very definite determined zones of the open ocean where food is particularly abundant. If the reflecting layer is eventually proved to be composed of fish, the prevailing views of fish distribution will have to be radically revised.

The most startling theory (and the one that seems to have the fewest supporters) is that the layer consists of concentrations of squid, 'hovering below the illuminated zone of the sea and awaiting the arrival of darkness in which to resume their raids into the plankton-rich surface waters.' Proponents of this theory argue that squid are abundant enough, and of wide enough distribution, to give the echoes that have been picked up almost everywhere from the equator to the two poles. Squid are known to be the sole food of the sperm whale, found in the open oceans in all temperate and tropical waters. They also form the exclusive diet of the bottlenosed whale and are eaten extensively by most other toothed whales, by seals, and by many sea birds. All these facts argue that they must be prodigiously abundant.

It is true that men who have worked close to the sea surface at night have received vivid impressions of the abundance and activity of squids in the surface waters in darkness. Long ago Johan Hjort wrote:

"One night we were hauling long lines on the Faroe slope, working with an electric lamp hanging over the side in order to see the line, when like lightning flashes one squid after another shot towards the light . . . In October 1902 we were one night steaming outside the slopes of the coast banks of Norway, and for many miles we could see the squids moving in the surface waters like luminous bubbles, resembling large milky white electric lamps being constantly lit and extinguished."

Thor Heyerdahl reports that at night his raft was literally bombarded by squids; and Richard Fleming says that in his oceanographic work off the coast of Panama it was common to see immense schools of squid gathering at the surface at night and leaping upward toward the lights that were used by the men to operate their instruments. But equally spectacular surface displays of shrimp have been seen, and most people find it difficult to believe in the ocean-wide abundance of squid.

•

H.G. Wells

The difference between what works and what is true is explored by novelist H.G. Wells (1866-1946) in his survey of the social sciences The Work, Wealth, and Happiness of Mankind.

I wish that there was a plain and popular book in existence upon the history of scientific ideas. It would be fascinating to reconstruct the intellectual atmosphere that surrounded Galileo and show the pre-existing foundations on which his ideas were based. Or ask what did Gilbert, the first student of magnetism, know, and what was the ideology with which the natural philosophers of the Stuart period had to struggle? It would be very interesting and illuminating to trace the rapid modification of these elementary concepts as the scientific process became vigorous and spread into general thought.

Few people realize how recent that invasion is, how new the current diagram of the universe is, and how recently the ideas of modern science have reached the commoner sort of people. The present writer is sixty-five. When he was a little boy his mother taught him out of a book she valued very highly, *Magnell's Questions*. It had been her own school book. It was already old-fashioned, but it was still in use and on sale. It was a book on the eighteenth-century plan of question and answer, and it taught that there were four elements, earth, air, fire, and water.

These four elements are as old at least as Aristotle. It never occurred to me in my white-sock and plaid-petticoat days to ask in what proportion these fundamental ingredients were mixed in myself or the tablecloth or my bread and milk. I just swallowed them as I swallowed the bread and milk.

From Aristotle I made a stride to the eighteenth century. The two elements of the Arabian alchemists, sulphur and mercury, I never heard of then, nor of Paracelsus and his universe of salt, sulphur, mercury, water, and the vital elixir. None of that ever got through to me. I went to a boys' school, and there I learnt, straightaway, that I was made up of hard, definite molecules, built up of hard, definite, indestructible atoms of carbon, oxygen, hydrogen, nitrogen, phosphorus, calcium, sodium, chlorine, and a few others. These were the real elements. They were shown plainly in my textbook like peas or common balls suitably grouped. That also I accepted for a time without making any fuss about it. I do not remember parting with the Four Elements: they got lost and I went on with the new lot.

At another school, and then at the Royal College of Science I learnt of a simple eternity of atoms and force. But the atoms now began to be less solid and simple. We talked very much of ether and protyle at the Royal College, but protons and electrons were still to come, and atoms, though taking on strange shapes and movements, were intact. Atoms could neither be transformed nor destroyed, but forces, though they could not be destroyed, could be transformed. This indestructibility of the chameleon of force was the celebrated Conservation of Energy, which has since lost prestige, though it remains as a sound working generalization for the everyday engineer.

But in those days, when I debated and philosophized with my fellow students, I was speedily made aware that these atoms and molecules were not realities at all; they were, it was explained to me, essentially mnemonics; they satisfied, in the simplest possible arrangement of material models and images, what was needed to assemble and reconcile the known phenomena of matter. That was all they were. That I grasped without much difficulty. There was no shock to me, therefore, when presently new observations necessitated fresh elaborations of the model. My schoolmaster had been a little too crude in his instructions. He had not been a scientific man, but only a teacher of science. He had been an unredeemed Realist, teaching science in a dogmatic Realist way. Science, I now understood, never contradicts herself absolutely, but she is always busy in revising her classifications and touching up and rephrasing her earlier cruder statements. Science never professes to present more than a working diagram of fact. She does not *explain*, she *states the relations and associations of facts as simply as possible.*

Her justification for her diagrams lies in her increasing power to change matter. The test of all her theories is that they work. She has always been true, and continually she becomes truer. But she never expects to reach Ultimate Truth. At their truest her theories are not, and never pretend to be, more than diagrams to fit, not even all possible facts, but simply the known facts.

In my student days, forty-five years ago now, we were already quite aware that the *exact* equivalence of cause and effect was no more than a convenient convention, and that it was possible to represent the universe as a system of unique events in a spacetime framework. These are not new ideas. They were then common student talk. When excited journalists announce that such intellectualists as Professor Eddington and Professor Whitehead have made astounding discoveries to overthrow the "dogmas of science," they are writing in sublime ignorance of the fact that there are no dogmas of science and that these ideas that seem such marvellous "discoveries" to them have been in circulation for more than half a century.

No engineer bothers about these considerations of marginal error and the relativity of things, when he plans out the making of a number of machines "in series" with replaceable parts. Every part is unique indeed and a little out of the straight, but it is near enough and straight enough to serve. The machines work. And no appreciable effect has been produced upon the teaching of machine drawing by the possibility that space is curved and expanding. In this book, let the reader bear in mind, we are always down at the level of the engineer and the machine drawing. From cover to cover we are dealing with practical things on the surface of the earth, where gravitation is best represented as a centripetal pull, and where a pound of feathers weighs equal to a pound of lead, and things are what they seem. We deal with the daily life of human beings now and in the ages immediately ahead. We remain in the space and time of ordinary experience throughout this book, at an infinite distance from ultimate truth.

•

Sigmund Freud

Here is the first allusion of Sigmund Freud (1856-1939) to the "Oedipus Complex," from the chapter "Dreams of the Death of Beloved Persons" in The Interpretation of Dreams.

In my experience, which is already extensive, the chief part in the mental lives of all children who later become psychoneurotics is played by their parents. Being in love with the one parent and hating the other are among the essential constituents of the stock of psychical impulses which is formed at that time and which is of such importance in determining the symptoms of the later neurosis. It is not my belief, however, that psychoneurotics differ sharply in this respect from other human beings who remain normal — that they are able, that is, to create something absolutely new and peculiar to themselves. It is far more probable — and this is confirmed by occasional observations of normal children — that they are only distinguished by exhibiting on a magnified scale feelings of love and hatred to their parents which occur less obviously and less intensely in the minds of most children.

This discovery is confirmed by a legend that has come down to us from classical antiquity: a legend whose profound and universal power to move can only be understood if the hypothesis I have put forward in regard to the psychology of children has an equally universal validity. What I have in mind is the legend of King Oedipus and Sophocles' drama which bears his name.

Oedipus, son of Laius, King of Thebes, and of Jocasta, was exposed as an infant because an oracle had warned Laius that the still unborn child would be his father's murderer. The child was rescued, and grew up as a prince in an alien court, until, in doubts as to his origin, he too questioned the oracle and was warned to avoid his home since he was destined to murder his father and take his mother in marriage. On the road leading away from what he believed was his home, he met King Laius and slew him in a sudden quarrel. He came next to Thebes and solved the riddle set him by the Sphinx who barred his way. Out of gratitude the Thebans made him their king and gave him Jocasta's hand in marriage. He reigned long in peace and honor, and she who, unknown to him, was his mother bore him two sons and two daughters. Then at last a plague broke out and the Thebans made enquiry once more of the oracle. It is at this point that Sophocles' tragedy opens. The messengers bring back the reply that the plague will cease when the murderer of Laius has been driven from the land.

> But he, where is he? Where shall now be read
> The fading record of this ancient guilt?

The action of the play consists in nothing other than the process of revealing, with cunning delays and ever-mounting excitement—a process that can be likened to the work of a psychoanalysis—that Oedipus himself is the murderer of Laius, but further that he is the son of the murdered man and of Jocasta. Appalled at the abomination which he has unwittingly perpetrated, Oedipus blinds himself and forsakes his home. The oracle has been fulfilled.

Oedipus Rex is what is known as a tragedy of destiny. Its tragic effect is said to lie in the contrast between the supreme will of the gods and the vain attempts of mankind to escape the evil that threatens them. The lesson which, it is said, the deeply moved spectator should learn from the tragedy is submission to the divine will and realization of his own impotence. Modern dramatists have accordingly tried to achieve a similar tragic effect by weaving the same contrast into a plot invented by themselves. But the spectators have looked on unmoved while a curse or an oracle was fulfilled in spite of all the efforts of some innocent man: later tragedies of destiny have failed in their effect.

If *Oedipus Rex* moves a modern audience no less than it did the contemporary Greek one, the explanation can only be that its effect does not lie in the contrast between destiny and human will, but is to be looked for in the particular nature of the material on which that contrast is exemplified. There must be something which makes a voice within us ready to recognize the compelling force of destiny in the *Oedipus*, while we can dismiss as merely arbitrary such dispositions as are laid down in [Grillparzer's] *Die Ahnfrau* or other modern tragedies of destiny. And a factor of this kind is in fact involved in the story of King Oedipus. His destiny moves us only because it might have been ours—because the oracle laid the same curse upon us before our birth as upon him. It is the fate of all of us, perhaps, to direct our first sexual impulse towards our mother and our first hatred and our first murderous wish against our father. Our dreams convince us that that is so. King Oedipus, who slew his father Laius and married his mother Jocasta, merely shows us the fulfillment of our own childhood wishes. But, more fortunate than he, we have meanwhile succeeded, in so far as we have not become psychoneurotics, in detaching our sexual impulses from our mothers and in forgetting our jealousy of our fathers. Here is one in whom these primaeval wishes of our childhood have been fulfilled, and we shrink back from him with the whole force of the repression by which those wishes have since that time been held down within us. While the poet, as he unravels the past, brings to light the guilt of Oedipus, he is at the same time compelling us to recognize our own inner minds, in which those same impulses, though suppressed, are still to be found. The contrast with which the closing Chorus leaves us confronted—

> . . . Fix on Oedipus your eyes,
> Who resolved the dark enigma, noblest champion and most wise.
> Like a star his envied fortune mounted beaming far and wide:
> Now he sinks in seas of anguish, whelmed beneath a raging tide. . .

—strikes as a warning at ourselves and our pride, at us who since our childhood have grown so wise and so mighty in our own eyes. Like Oedipus, we live in ignorance of these wishes, repugnant to morality, which have been forced upon us by Nature, and after their revelation we may all of us well seek to close our eyes to the scenes of our childhood.

•

Bertrand Russell

Bertrand Russell (1872-1970) wrote an essay in the New York Times Magazine entitled "The Science to Save Us from Science," from which the following excerpt is taken. It is an ideal essay to begin a discussion of ideals in science.

I call an ideal "intelligent" when it is possible to approximate to it by pursuing it. This is by no means sufficient as an ethical criterion, but it *is* a test by which many aims can be condemned. It cannot be supposed that Hitler desired the fate which he brought upon his country and himself, and yet it was pretty certain that this would be the result of his arrogance. Therefore the ideal of "Deutschland ueber Alles" can be condemned as unintelligent. (I do not mean to suggest that this is its only defect.) Spain, France, Germany, and Russia have successively sought world dominion: three of them have endured defeat in consequence, but their fate has not inspired wisdom.

Whether science — and indeed civilization in general — can long survive depends upon psychology, that is to say, it depends upon what human beings desire. The human beings concerned are rulers in totalitarian countries, and the mass of men and women in democracies. Political passions determine political conduct much more directly than is often supposed. If men desire victory more than cooperation, they will think victory possible.

But if hatred so dominates them that they are more anxious to see their enemies killed than to keep their own children alive, they will discover all kinds of "noble" reasons in favor of war. If they resent inferiority or wish to preserve superiority, they will have the sentiments that promote the class war. If they are bored beyond a point, they will welcome excitement even of a painful kind.

Such sentiments, when widespread, determine the policies and decisions of nations. Science can, if rulers so desire, create sentiments which will avert disaster and facilitate cooperation. At present there are powerful rulers who have no such wish. But the possibility exists, and science can be just as potent for good as for evil. It is not science, however, which will determine how science is used.

Science, by itself, cannot supply us with an ethic. It can show us how to achieve a given end, and it may show us that some ends cannot be achieved. But among ends that can be achieved our choice must be decided by other than purely scientific considerations. If a man were to say, "I hate the human race, and I think it would be a good thing if it were exterminated," we could say, "Well, my dear sir, let us begin the process with you." But this is hardly argument, and no amount of science could prove such a man mistaken.

But all who are not lunatics are agreed about certain things: That it is better to be alive than dead, better to be adequately fed than starved, better to be free than a slave. Many people desire those things only for themselves and their friends; they are quite content that their enemies should suffer. These people can be refuted by science: Mankind has become so much one family that we cannot insure our own prosperity except by insuring that of everyone else. If you wish to be happy yourself, you must resign yourself to seeing others also happy.

Whether science can continue, and whether, while it continues, it can do more good than harm, depends upon the capacity of mankind to learn this simple lesson. Perhaps it is necessary that all should learn it, but it must be learned by all who have great power, and among those some still have a long way to go.

•

Albert Einstein

Along with Darwin's theories of evolution and Freud's discovery that dreams could divulge accurate psychological information about the dreamer, the other great breakthrough in scientific thought within the last century was Albert Einstein's theory of relativity. Here is the entirety of a brief article written by him in 1946 for a popular science magazine.

$E = mc^2$
In order to understand the law of the equivalence of mass and energy, we must go back to two conservation or "balance" principles which, independent of each other, held a high place in pre-relativity physics. These were the principle of the conservation of energy and the principle of the conservation of mass. The first of these, advanced by Leibnitz as long ago as the seventeenth century, was developed in the nineteenth century essentially as a corollary of a principle of mechanics.

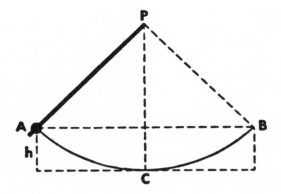

Consider, for example, a pendulum whose mass swings back and forth between the points A and B. At these points the mass m is higher by the amount h than it is at C, the lowest point of the path (see drawing). At C, on the other hand, the lifting height has disappeared and instead of it the mass has a velocity v. It is as though the lifting height could be converted entirely into velocity, and vice versa. The exact relation would be expressed as $mgh = \frac{m}{2}v^2$, with g representing the acceleration of gravity. What is interesting here is that this relation is independent of both the length of the pendulum and the form of the path through which the mass moves.

The significance is that something remains constant throughout the process, and that something is energy. At A and at B it is an energy of position, or "potential" energy; at C it is an energy of motion, or "kinetic" energy. If this concept is correct, then the sum $mgh + m\frac{v^2}{2}$ must have the same value for any position of the pendulum, if h is understood to represent the height above C, and v the velocity at that point in the pendulum's path. And such is found to be actually the case. The generalization of this principle gives us the law of the conservation of mechanical energy. But what happens when friction stops the pendulum?

The answer to that was found in the study of heat phenomena. This study, based on the assumption that heat is an indestructible substance which flows from a warmer to a colder object, seemed to give us a principle of the "conservation of heat." On the other hand, from time immemorial it has been known that heat could be produced by friction, as in the fire-making drills of the Indians. The physicists were for long unable to account for this kind of heat "production." Their difficulties were overcome only when it was successfully established that, for any given amount of heat produced by friction, an exactly proportional amount of energy had to be expended. Thus did we arrive at a principle of the "equivalence of work and heat." With our pendulum, for example, mechanical energy is gradually converted by friction into heat.

In such fashion the principles of the conservation of mechanical and thermal energies were merged into one. The physicists were thereupon persuaded that the conservation principle could be further extended to take in chemical and electromagnetic processes—in short, could be applied to all fields. It appeared that in our physical system there was a sum total of energies that remained constant through all changes that might occur.

Now for the principle of the conservation of mass. Mass is defined by the resistance that a body opposes to its acceleration (inert mass). It is also measured by the weight of the body (heavy mass). That these two radically different definitions lead to the same value for the mass of a body is, in itself, an astonishing fact. According to the principle—namely, that masses remain unchanged under any physical or chemical changes—the mass appeared to be the essential (because unvarying) quality of matter. Heating, melting, vaporization, or combining into chemical compounds would not change the total mass.

Physicists accepted this principle up to a few decades ago. But it proved inadequate in the face of the special theory of relativity. It was therefore merged with the energy principle — just as, about 60 years before, the principle of the conservation of mechanical energy had been combined with the principle of the conservation of heat. We might say that the principle of the conservation of energy, having previously swallowed up that of the conservation of heat, now proceeded to swallow that of conservation of mass — and holds the field alone.

It is customary to express the equivalence of mass and energy (though somewhat inexactly) by the formula $E = mc^2$, in which c represents the velocity of light, about 186,000 miles per second. E is the energy that is contained in a stationary body; m is its mass. The energy that belongs to the mass m is equal to this mass, multiplied by the square of the enormous speed of light—which is to say, a vast amount of energy for every unit of mass.

But if every gram of material contains this tremendous energy, why did it go so long unnoticed? The answer is simple enough: so long as none of the energy is given off externally, it cannot be observed. It is as though a man who is fabulously rich should never spend or give away a cent; no one could tell how rich he was.

Now we can reverse the relation and say that an increase of E in the amount of energy must be accompanied by an increase of $\frac{m}{c^2}$ in the mass.

I can easily supply energy to the mass—for instance, if I heat it by 10 degrees. So why not measure the mass increase, or weight increase, connected with this change? The trouble here is that in the mass increase the enormous factor c^2 occurs in the denominator of the fraction. In such a case the increase is too small to be measured directly; even with the most sensitive balance.

For a mass increase to be measurable, the change of energy per mass unit must be enormously large. We know of only one sphere in which such amounts of energy per mass unit are released: namely, radioactive disintegration. Schematically, the process goes like this: an atom of the mass M splits into two atoms of the mass M' and M", which separate with tremendous kinetic energy. If we imagine these two masses as brought to rest—that is, if we take this energy of motion from them—then, considered together, they are essentially poorer in energy than was the original atom. According to the equivalence principle, the mass sum M' + M" of the disintegration products must also be somewhat smaller than the original mass M of the disintegrating atom—in contradiction to the old principle of the conservation of mass. The relative difference of the two is on the order of 1/10 of one percent.

Now, we cannot actually weigh the atoms individually. However, there are indirect methods for measuring their weights exactly. We can likewise determine the kinetic energies that are transferred to the disintegration products M' and M". Thus it has become possible to test and confirm the equivalence formula. Also, the law permits us to calculate in advance, from precisely determined atom weights, just how much energy will be released with any atom disintegration we have in mind. The law says nothing, of course, as to whether—or how—the disintegration reaction can be brought about.

What takes place can be illustrated with the help of our rich man. The atom M is a rich miser who, during his life, gives away no money *(energy)*. But in his will he bequeaths his fortune to his sons M' and M", on condition that they give to the community a small amount, less than one thousandth of the whole estate *(energy or mass)*. The sons together have somewhat less than the father had *(the mass sum M' + M" is somewhat smaller than the mass M of the radioactive atom)*. But the part given to the community, though relatively small, is still so enormously large *(considered as kinetic energy)* that it brings with it a great threat of evil. Averting that threat has become the most urgent problem of our time.

•

Robert Louis Stevenson

In this passage from Robert Louis Stevenson's Dr. Jekyll and Mr. Hyde (published in 1886), Dr. Lanyon becomes the first witness to the effects gone awry of the scientific experiment of Dr. Jekyll. The power of science to release and empower the demonic, evil nature of man has been a theme in many popular works of fiction for several centuries, but none exceed this one for dramatic effect.

"I beg your pardon, Dr. Lanyon," he replied civilly enough. "What you say is very well founded; and my impatience has shown its heels to my politeness. I come here at the instance of your colleague, Dr. Henry Jekyll, on a piece of business of some moment; and I understood..." He paused and put his hand to his throat, and I could see, in spite of his collected manner, that he was wrestling against the approaches of the hysteria—"I understood, a drawer..."

But here I took pity on my visitor's suspense, and some perhaps on my own growing curiosity.

"There it is, sir," said I, pointing to the drawer, where it lay on the floor behind a table and still covered with the sheet.

He sprang to it, and then paused, and laid his hand upon his heart: I could hear his teeth grate with the convulsive action of his jaws; and his face was so ghastly to see that I grew alarmed both for his life and reason.

"Compose yourself," said I.

He turned a dreadful smile to me, and as if with the decision of despair, plucked away the sheet. At sight of the contents, he uttered one loud sob of such immense relief that I sat petrified. And the next moment, in a voice that was already fairly well under control, "Have you a graduated glass?" he asked.

I rose from my place with something of an effort and gave him what he asked.

He thanked me with a smiling nod, measured out a few minims of the red tincture and added one of the powders. The mixture, which was at first of a reddish hue, began, in proportion as the crystals melted, to brighten in color, to effervesce audibly, and to throw off small fumes of vapor. Suddenly and at the same moment, the ebullition ceased and the compound changed to a dark purple, which faded again more slowly to a watery green. My visitor, who had watched these metamorphoses with a keen eye, smiled, set down the glass upon the table, and then turned and looked upon me with an air of scrutiny.

"And now," said he, "to settle what remains. Will you be wise? will you be guided? will you suffer me to take this glass in my hand and to go forth from your house without further parley? or has the greed of curiosity too much command of you? Think before you answer, for it shall be done as you decide. As you decide, you shall be left as you were before, and neither richer nor wiser, unless the sense of service rendered to a man in mortal distress may be counted as a kind of riches of the soul. Or, if you shall so prefer to choose, a new province of knowledge and new avenues to fame and power shall be laid open to you, here, in this room, upon the instant; and your sight shall be blasted by a prodigy to stagger the unbelief of Satan."

"Sir," said I, affecting a coolness that I was far from truly possessing, "you speak enigmas, and you will perhaps not wonder that I hear you with no very strong impression of belief. But I have gone too far in the way of inexplicable services to pause before I see the end."

"It is well," replied my visitor. "Lanyon, you remember your vows: what follows is under the seal of our profession. And now, you who have so long been bound to the most narrow and material views, you who have denied the virtue of transcendental medicine, you who have derided your superiors—behold!"

He put the glass to his lips and drank at one gulp. A cry followed; he reeled, staggered, clutched at the table and held on, staring with injected eyes, gasping with open mouth; and as I looked there came, I thought, a change—he seemed to swell—his face became suddenly black and the features seemed to melt and alter—and the next moment, I had sprung to my feet and leaped back against the wall, my arm raised to shield me from that prodigy, my mind submerged in terror.

"O God!" I screamed, and "O God!" again and again; for there before my eyes—pale and shaken, and half fainting, and groping before him with his hands, like a man restored from death — there stood Henry Jekyll!

What he told me in the next hour, I cannot bring my mind to set on paper. I saw what I saw, I heard what I heard, and my soul sickened at it; and yet now when that sight has faded from my eyes, I ask myself if I believe it, and I cannot answer. My life is shaken to its roots; sleep has left me; the deadliest terror sits by me at all hours of the day and night; and I feel that my days are numbered, and that I must die; and yet I shall die incredulous. As for the moral turpitude that man unveiled to me, even with tears of penitence, I cannot, even in memory, dwell on it without a start of horror. I will say but one thing, Utterson, and that (if you can bring your mind to credit it) will be more than enough. The creature who crept into my house that night was, on Jekyll's own confession, known by the name of Hyde and hunted for in every corner of the land as the murderer of Carew.

•

Bernard Cooper

Bernard Cooper, a teacher of literature at Otis/Parsons in Los Angeles, has combined his ideas of the termination of the dinosaurs with impressions at the gym to create this whimsical piece of creative writing, called "Sudden Extinction," published in the Winter, 1988 issue of the literary journal Grand Street.

Sudden Extinction

The vertebrae of dinosaurs, found in countless excavations, are dusted and rinsed and catalogued. We guess and guess at their huge habits as we gaze at the fossils which capture their absence, sprawling three-toed indentations, the shadowy lattice of ribs. Their skulls are a slight embarrassment, snug even for a head full of blunt wants and backward motives. The brachiosaurus's brain, for example, sat atop his tapered neck like a minuscule flame on a mammoth candle.

My favorite is triceratops, his face a hideous Rorschach blot of broad bone and blue hide. The Museum of Natural History owns a replica that doesn't do him justice. One front foot is poised in the air like an elephant sedated for a sideshow. And the nasal horn for tossing aggressors is as dull and mundane as a hook for a hat.

One prominent paleontologist believes that during an instantaneous ice age, glaciers encased these monsters mid-meal — stegosaurus, podokesaurus, iguanodon — all trapped forever like spectrums in glass. But suppose extinction was a matter of choice, and they just didn't want to stand up any longer, like drunk guests at a party's end who pass out in the dark den. There are guys at my gym whose latissimus dorsi, having spread like thunderheads, cause them to inch through an ordinary door; might the dinosauria have grown too big of their own volition?

Derek speaks in expletives and swears that one day his back will be as big as a condominium. Mike's muscles, marbled with veins, perspire from ferocious motion, the taut skin about to split. When Bill does a bench press, the barbell bends from fifty-pound plates; his cheeks expand and expel great gusts of spittle and air till his face and eyes are flushed with blood and his elbows quiver, the weight sways, and someone runs over and hovers above him roaring for one more repetition.

Once, I imagined our exercise through x-ray eyes. Our skeletons gaped at their own reflection. Empty eyes, like apertures, opened onto an afterlife. Lightning-bright spines flashed from sacra. Phalanges of hands were splayed in surprise. Bones were glowing everywhere, years scoured down to marrow, flesh redressed with white.

And I knew our remains were meant to keep like secrets under the earth. And I knew one day we would topple like monuments, stirring up clouds of dust. And I almost heard the dirge of our perishing, thud after thud after thud, our last titanic exhalations long and labored and loud.

·

James Gleick

In his book Chaos, Making a New Science, *James Gleick calls upon history, philosophy, and literature to help describe the revolutionary birth of a new science that fell between the conventional delineation of the disciplines and defied "the traditional view that science progresses by the accretion of knowledge."*

The historian of science Thomas S. Kuhn describes a disturbing experiment conducted by a pair of psychologists in the 1940s. Subjects were given glimpses of playing cards, one at a time, and asked to name them. There was a trick, of course. A few of the cards were freakish: for example, a red six of spades or a black queen of diamonds.

At high speed the subjects sailed smoothly along. Nothing could have been simpler. They didn't see the anomalies at all. Shown a red six of spades, they would sing out either "six of hearts" or "six of spades." But when the cards were displayed for longer intervals, the subjects started to hesitate. They became aware of a problem but were not sure quite what it was. A subject might say that he had seen something odd, like a red border around a black heart.

Eventually, as the pace was slowed even more, most subjects would catch on. They would see the wrong cards and make the mental shift necessary to play the game without error. Not everyone, though. A few suffered a sense of disorientation that brought real pain. "I can't make that suit out, whatever it is," said one. "It didn't even look like a card that time. I don't know what color it is now or whether it's a spade or a heart. I'm not even sure what a spade looks like. My God!"

Professional scientists, given brief, uncertain glimpses of nature's workings, are no less vulnerable to anguish and confusion when they come face to face with incongruity. And incongruity, when it changes the way a scientist sees, makes possible the most important advances. So Kuhn argues, and so the story of chaos suggests.

Kuhn's notions of how scientists work and how revolutions occur drew as much hostility as admiration when he first published them, in 1962, and the controversy has never ended. He pushed a sharp needle into the traditional view that science progresses by the accretion of knowledge, each discovery adding to the last, and that new theories emerge when new experimental facts require them. He deflated the view of science as an orderly process of asking questions and finding their answers. He emphasized a contrast between the bulk of what scientists do, working on legitimate, well-understood problems within their disciplines, and the exceptional, unorthodox work that creates revolutions. Not by accident, he made scientists seem less than perfect rationalists.

In Kuhn's scheme, normal science consists largely of mopping-up operations. Experimentalists carry out modified versions of experiments that have been carried out many times before. Theorists add a brick here, reshape a cornice there, in a wall of theory. It could hardly be otherwise. If all scientists had to begin from the beginning, questioning fundamental assumptions, they would be hard pressed to reach the level of technical sophistication necessary to do useful work. In Benjamin Franklin's time, the handful of scientists trying to understand electricity could choose their own first principles—indeed, had to. One researcher might consider attraction to be the most important electrical effect, thinking of electricity as a sort of "effluvium" emanating from substances. Another might think of electricity as a fluid, conveyed by conducting material. These scientists could speak almost as easily to laymen as to each other, because they had not yet reached a stage where they could take for granted a common, specialized language for the phenomena they were studying. By contrast, a twentieth-century fluid dynamicist could hardly expect to advance knowledge in his field without first adopting a body of terminology and mathematical technique. In return, unconsciously, he would give up much freedom to question the foundations of his science.

Central to Kuhn's ideas is the vision of normal science as solving problems, the kinds of problems that students learn the first time they open their textbooks. Such problems define an accepted style of achievement that carries most scientists through graduate school, through their thesis work, and through the writing of journal articles that makes up the body of academic careers. "Under normal conditions the research scientist is not an innovator but a solver of puzzles, and the puzzles upon which he concentrates are just those which he believes can be both stated and solved within the existing scientific tradition," Kuhn wrote.

Then there are revolutions. A new science arises out of one that has reached a dead end. Often a revolution has an interdisciplinary character —its central discoveries often come from people straying outside the normal bounds of their specialties. The problems that obsess these theorists are not recognized as legitimate lines of inquiry. Thesis proposals are turned down or articles are refused publication. The theorists themselves are not sure whether they would recognize an answer if they saw one. They accept risk to their careers. A few freethinkers working alone, unable to explain where they are heading, afraid even to tell their colleagues what they are doing—that romantic image lies at the heart of Kuhn's scheme, and it has occurred in real life, time and time again, in the exploration of chaos.

Every scientist who turned to chaos early had a story to tell of discouragement or open hostility. Graduate students were warned that their careers could be jeopardized if they wrote theses in an untested discipline, in which their advisors had no expertise. A particle physicist, hearing about this new mathematics, might begin playing with it on his own, thinking it was a beautiful thing, both beautiful and hard—but would feel that he could never tell his colleagues about it. Older professors felt they were suffering a kind of midlife crisis, gambling on a line of research that many colleagues were likely to misunderstand or resent. But they also felt an intellectual excitement that comes with the truly new. Even outsiders felt it, those who were attuned to it. To Freeman Dyson at the Institute for Advanced Study, the news of chaos came "like an electric shock" in the 1970s. Others felt that for the first time in their professional lives they were witnessing a true paradigm shift, a transformation in a way of thinking.

•

William Shakespeare

Here is a view of science from William Shakespeare (1564-1616), from Troilus and Cressida.

The heavens themselves, the planets, and this centre,
Observe degree, priority, and place,
Insisture, course, proportion, season, form,
Office, and custom, in all line of order:

And therefore is the glorious planet Sol
In noble eminence enthroned and sphered
Amidst the other; whose med'cinable eye
Corrects the ill aspects of planets evil,
And posts, like the commandment of a king,
Sans check, to good and bad: but when the planets,
In evil mixture, to disorder wander,
What plagues, and what portents, what mutiny,
What raging of the sea, shaking of earth,
Commotion in the winds, frights, changes, horrors,
Divert and crack, rend and deracinate
The unity and married calm of states
Quite from their fixure! O, when degree is shaked,
Which is the ladder to all high designs,
The enterprise is sick! How could communities,
Degrees in schools, and brotherhoods in cities,
Peaceful commerce from dividable shores,
The primogenity and due of birth,
Prerogative of age, crowns, sceptres, laurels,
But by degree, stand in authentic place?
Take but degree away, untune that string,
And, hark, what discord follows! each thing meets
In mere oppugnancy: the bounded waters
Should lift their bosoms higher than the shores,
And make a sop of all this solid globe:
Strength should be lord of imbecility,
And the rude son should strike his father dead:
Force should be right; or rather, right and wrong—
Between whose endless jar justice resides—
Should lose their names, and so should justice too.
Then every thing includes itself in power,
Power into will, will into appetite.
And appetite, an universal wolf,
So double seconded with will and power,
Must make perforce an universal prey,
And last eat up himself.

•

Bentley Glass

In his 1985 book Progress or Catastrophe, *biologist Bentley Glass explores the nature of biological science and its impact on society. In the following excerpts he discusses the relationship between indeterminacy in physics and biology. Applications of chaos theory are apparent here.*

The second reason for doubting the reducibility of all biological laws to those of physics and chemistry is in fact a simple one, analogous to, but by no means identical with, the principle of indeterminacy in physics. It derives from the nature of the evolutionary process which affects all life.

The principle of indeterminacy in physics relates to single quanta of energy and single elementary particles. There is no reason to try to introduce this principle into biology, as some thinkers—I believe foolishly—have tried to do, in order to explain will or consciousness. Whatever material events lie at the basis of mental phenomena, they seem surely to depend upon a multimolecular basis. Like the events of chemistry, they involve the statistical behavior of millions or billions of molecules. But there is nevertheless a significant analogy between indeterminacy in physics and that in biology. The indeterminacy in physics arises because a single quantum or elementary particle cannot behave statistically at a given instant. It is unique, and the very operations performed to determine its position in space and its energy unavoidably alter the one or the other. In evolutionary biology, where one is concerned with past events—the mutations and the selections—that have led to the existence of a certain contemporary pattern in the DNA of a particular species, one is likewise dealing with uniqueness, the particularity of history. [.....]

But the uniqueness of the particular event, embedded in the history and evolution of life, seems an unanswerable argument for the impossibility of explaining all aspects of life in terms of the laws of physical science which are demonstrable in nonliving systems. Finally, the randomness of behavior at distinct levels of organization leads to statistical laws based on probabilities that are applicable strictly to the entities that are exhibiting the random behavior. These statistical laws are explications as final at one level as at another. They relate to the nature of what we mean by "randomness," by "chance," and by "probability," rather than to the physical and chemical nature of material systems and measurable forces upon which our classical ideas of causation and determinism are founded. They find their unity in mathematics. They express the ultimate uncertainty of particular events in the real world where the laws of science deal only with the average outcomes.

•

Sally J. Kim

Sally J. Kim was a senior at Bronx High School of Science when she conducted a cell biology experiment at Cornell University Medical College, which proved to be a winner in the 1988 Westinghouse Science Talent Search contest. The abstract that follows—from her paper entitled "Immunochemical Analysis of an M-band Antigen in the Myofibrils of Skeletal Muscle"—uses such a specialized vocabulary that it sounds almost like another language.

Two monoclonal antibodies (CP-7 & CP-13) which were prepared against crude myosin from chicken pectoral muscle, were found to bind to the M-region of the sarcomere by immunofluorescence microscopy. The objective of this study was to determine the molecular weights of the two antigens detected by these antibodies and to characterize these proteins by immunochemical methods. The expression of the two antigens in chicken pectoral muscle was examined by Sodium Dodecyl Sulfate Polyacrylamide Gel Electrophoresis (SDS-PAGE), Western blots, immunofluorescence, enzyme-linked immunosorbent assays (ELISA), and electron microscopy. The immunoblots showed that the antigen recognized by CP-13 has a molecular weight of 185 kilodaltons (kD). For convenience the protein was termed MP-185 (M-line protein, 185 kD). Competitive binding ELISA experiments suggest that CP-7 and CP-13 recognize the same protein since they exhibit steric interference. Other ELISA experiments revealed that the expression of both antigens increased in parallel with myofibril development. Using immunofluorescence microscopy, CP-13 was shown to bind as a single line in the M-region. However, high resolution electron micrographs revealed that the antibody was localized to several striations in the central region of the sarcomere. Efforts were also made to affinity purify CP-13's antigen but inherent problems in technique led to a crude preparation of MP-185.

•

Samuel Taylor Coleridge

In the following section from his Religious Musings, *Samuel Taylor Coleridge (1772-1850) presents an exalted vision of the origins of science.*

From Avarice thus, from Luxury and War
Sprang heavenly Science, and from Science Freedom.
O'er waken'd realms Philosophers and Bards
Spread in concentric circles: they whose souls,
Conscious of their high dignities from God,
Brook not Wealth's rivalry! and they, who long
Enamoured with the charms of order, hate
The unseemly disproportion: and whoe'er
Turn with mild sorrow from the Victor's car
And the low puppetry of thrones, to muse
On that blest triumph, when the Patriot Sage
Called the red lightnings from the o'er-rushing cloud
And dashed the beauteous terrors on the earth
Smiling majestic. Such a phalanx ne'er
Measured firm paces to the calming sound
Of Spartan flute! These on the fated day,
When, stung to rage by Pity, eloquent men
Have roused with pealing voice the unnumbered tribes
That toil and groan and bleed, hungry and blind—
These hush'd awhile with patient eye serene,
Shall watch the mad careering of the storm;
Then o'er the wild and wavy chaos rush
And tame the outrageous mass, with plastic might
Moulding Confusion to such perfect forms,
As erst were wont,—bright visions of the day!—
To float before them, when, the summer noon,
Beneath some arched romantic rock reclined
They felt the sea-breeze lift their youthful locks;
Or in the month of blossoms, at mild eve,
Wandering with desultory feet inhaled
The wafted perfumes, and the flocks and woods
And many-tinted streams and setting sun
With all his gorgeous company of clouds
Ecstatic gazed! then homeward as they strayed
Cast the sad eye to earth, and inly mused
Why there was misery in a world so fair.

•

Dale Worsley

In this passage from Dale Worsley's play Blue Devils, *a science fiction fantasy in which Amelia Earhart and John James Audubon meet in the future, Amelia describes the combustion engine to John, who died before its invention. To her, it was more than a curiosity, it was a mechanism of liberation and adventure.*

AMELIA:

Picture a block of metal, yeah big. *(Indicates with hands.)* Drill a hole in it. A cylinder. *(Indicates size and placement of hole. Picks up a can of beans.)* Install a piston to move within the cylinder. *(Illustrates with can of beans.)* Spray an explosive mixture of gas and air into the cylinder. Compress it with the piston. Put a spark in. Pow. The explosion drives the piston down. Connect the piston to a shaft that turns. Connect the shaft to a wheel, or a propellor. And you have automobiles, powerboats, airplanes. Now. The four strokes of the Otto Cycle. I'll tell you who Otto was later. Let's go to the cylinder. First stroke: piston goes down, sucks in gas and air. Second stroke: piston goes up, compressing gas and air into a tight, volatile mixture, crammed with potential energy. Third stroke: Spark. Explosion. Bang. Piston shoots down to turn the crankshaft. This is the work. Speed, power, height, distance. Fourth stroke: piston goes back up, driving out exhausted fuel and air. And you're ready to start again. Rapidly. Line up, or put in a circle, two, or three, or four…six… eight…twelve cylinders. Get them all coordinated. Thoom. Thoom. Bdrdrdrdrd. Look out, it's getting hot. Cool it, surround it with a jacket of water, or run a breeze over it with a fan. The walls of the cylinder are about to fuse with friction. Keep them in a perpetual bath of oil. The exhaust could easily kill you. Run it down a pipe away from you. The noise is earsplitting. Muffle it. Get on this engine. Ride it. Take ten thousand people across the ocean on it. Fly it around the world.

•

Kenneth R. Manning

Ernest Everett Just (1883-1941) was a prominent American biologist who focused on the process of fertilization in marine invertebrates. As a black, he encountered racial prejudices that forced him to work in Europe. In a review of Black Apollo of Science, the Life of Ernest Everett Just, *Bentley Glass says, "Had Just been born in 1983 instead of 1883, or even in 1933, he might have risen to enjoy a position in one of our greater universities—as his biographer does (his biographer, Kenneth R. Manning, is also black)—and have come to greater fruition." Even so, Just was an admired and perhaps revolutionary biologist. In the following excerpt, Manning is referring to a conference in Princeton in 1935, the first Just attended after returning from an increasingly menacing Nazi Germany.*

Just gave two papers. The first, "Alcoholic Solutions as Gastric Secretogogues," was not the one he had prepared long and hard for, but it was important to him too. In it he demonstrated some of the possible medical relevance of his work. Foundations were likely to support research that could be shown to have practical applications in fields such as medicine and engineering. Perhaps, Just must have thought, he could bring this up with the Oberlaender Trust; perhaps he could even try to interest a specialized organization such as the Distilled Spirits Institute in Washington, D.C. Ten years earlier he had had no success in getting the National Research Council to take seriously his idea for protecting wharves from shipworm, but there was no harm in trying the practical route again, from a different angle and with a different organization. Also, it was good to show his Princeton audience that he was more than a tinkerer with marine eggs. He had never published or even talked about his shipworm work with the Woods Hole people, so for them this was a completely new angle.

The second paper, the big one, "Nuclear Increase during Development as a Factor in Differentiation and in Heredity," explained his theory of the gene and attempted to reconcile two opposing factions in the gene controversy, the embryologists and the geneticists. It was part of a larger paper, "A Single Theory for the Physiology of Development and Genetics," which was to be published a few months later at Just's own expense.

The talk at Princeton and the subsequent paper were an attack on the geneticist T. H. Morgan, who, though he had recently won the Nobel Prize for his work on *Drosophila*, had not come close to offering a gene theory that would satisfy an embryologist, especially Just. Just had spent time earlier with the Morgan group at Columbia and Woods Hole, but he had lost contact after his trips to Europe and no longer felt any loyalty in that direction.

At the time, a scientific controversy was raging about genetics and the physiology of development, or more accurately, about how those two fields could best be brought together. The embryologists were concerned with the development of the fertilized egg into differentiated adult characters, geneticists with the transmission of heredity elements from generation to generation. Both groups were dealing with concepts of heredity which related closely to the known distinction of genotype and phenotype. "Genotype" is the term used with respect to the genetic composition of an organism, that is, what genes it carries and passes on to the next generation. "Phenotype" is used to talk about how the organism looks, that is, its visible adult characters. The distinction had been made in 1911 by Wilhelm Johannsen in an article published in the *American Naturalist.* Just, though he did not use those terms, was attempting to show the interdependence of both phenomena. In other words, Just thought that "the same mode by which the grosser parts become differentiated calls forth the Mendelian differences."

Just admitted that there was no *a priori* reason to see development and heredity as the same, but he was prepared to show them both as different sides of the same coin, "merely two aspects of life-history." Morgan, the leader of the genetics group and himself a one-time embryologist, had tried to reconcile the two fields. But his book on the subject, published in 1934, treated the two disciplines separately, and by his own admission there was no synthesis to speak of. Curiously, Just never mentioned Morgan's book in either his talk or his article. As far as he was concerned, the problem remained—embryologists and geneticists were still not talking the same language.

The issue of differentiation was at the heart of the problem. The gene theorists, according to Just, postulated that "the genes order the progressive differentiation of development by liberating to the cytoplasm at different stages a different something which brings about differentiation." Just dismissed this as an explanation of the physiology of development. He also argued against the then popular theory known as embryonic segregation, held by embryologists such as Lillie and Conklin. He raised a number of objections to the theory and thereby put himself between the geneticists and the embryologists. He would join neither.

Just offered his own theory—a theory of restriction. He had arrived at it by observing the egg cell during cleavage and, by careful experimental work, noting the increase of the nucleus during this stage. He assumed that his observations of differentiation during cleavage were true for all other stages of the embryological process. The nuclear increase had to come from somewhere, he reasoned, since the nucleus could not build something out of nothing. Before identifying the source of the nuclear increase, Just further assumed that what is true of the nucleus is true of its constituents, the chromosomes. He cited direct indications, however, work that had been done on determining the presence of nucleic acid in eggs, that the chromosomes increased in mass. This building up of nuclei and chromosomes, which always went with differentiation, was done at the expense of the cytoplasm.

Just used his theory to explain numerous embryological concepts, from polyembryony to asexual reproduction. But he was not willing to stop there. He forged on, using his theory to place, somewhat incorrectly as it turned out, the basis of inheritance in the egg cytoplasm. His conception differs from the gene theory in two chief respects: by locating the factors for heredity in the cytoplasm instead of in the genes, and by offering "an interpretation of 'how the genes in the chromosomes produce the effects in and through the cytoplasm' of which Morgan confesses ignorance."

Biochemistry was becoming more and more important for Just and especially for his work on the nuclear increase during cleavage. In the mid-1930s no one knew the precise composition of the genetic substance, although by then it was becoming generally accepted that the nucleus consists essentially of nucleoprotein. But Just's mind was open to other possibilities. He called for further biochemical research into "the role of each of the three organic compounds in protoplasm." His work on nuclear increase pointed to that need. Only once (later, in 1940) did Just actually speculate rather offhandedly that "the genes are nucleic acids." He was never tied down to a belief in the primacy of the nucleus, as he knew some protoplasmic systems that had no discrete nucleus as such. Some bacteria and blue-green algae, he remembered, were nonnuclear organisms, containing mostly nucleic acid. But what he wanted was a theory of heredity applicable to all animals and all plants. He asked, then, how "upon the thread of the nucleic acid common to plants and that to animals, the specific gene entities are imposed." Biochemists would need to work overtime to provide the answer.

Just's theory of genetic restriction was an attempt to provide an explanation of the mechanism of differentiation. Never mind that the problem of differentiation was, and still is, a difficult one; exactly how it occurs continues to elude embryologists. In Just's view his theory had several advantages; perhaps most important, it gave the nucleus, cytoplasm, and ectopolasm definite roles in development—something that other theories had not done.

His "interpretation," as he called it, was recognized as imaginative and provocative. Younger scientists, Lester Barth among them, walked away from the lecture in awe. They wrote for reprints of the paper and read it with deep interest. Still, these scientists did not carry on and develop this work in any real way, even though they saw Just as one of the most creative men in zoology in the United States, someone who had dared to advance a new idea, someone who was different from the older scientists but showed no fear of them, someone who should be taken seriously and given equal respect. Biology was perhaps changing, Just thought, thanks to this new younger breed—which would eventually discover the real magic of the gene. Just hoped so, and if he did nothing else, he wanted to spur them on.

•

Gerard Manley Hopkins

In this poem, Gerard Manley Hopkins (1844-89) compares himself to a comet. Hopkins, a meditative Jesuit, was also a keen observer of nature. His journals are filled with wonderful descriptions and drawings of his observations.

    —I am like a slip of comet,
Scarce worth discovery, in some corner seen
Bridging the slender difference of two stars,
Come out of space, or suddenly engender'd
By heady elements, for no man knows;
But when she sights the sun she grows and sizes
And spins her skirts out, while her central star
Shakes its cocooning mists; and so she comes
To fields of light; millions of travelling rays
Pierce her; she hangs upon the flame-cased sun,
And sucks the light as full as Gideon's fleece:
But then her tether calls her; she falls off,
And as she dwindles shreds her smock of gold
Between the sistering planets, till she comes
To single Saturn, last and solitary;
And then she goes out into the cavernous dark.
So I go out: my little sweet is done:
I have drawn heat from this contagious sun:
To not ungentle death now forth I run.

•

Annie Dillard

Annie Dillard, a contributing editor to Harper's *magazine and a columnist for the* Wilderness Society *when she wrote her book* Pilgrim at Tinker Creek, *is a naturalist with a poetic bent and mystical temperament. In the following passage we are party to the close observation of a pet fish.*

A rosy, complex light fills my kitchen at the end of these lengthening June days. From an explosion on a nearby star eight minutes ago, the light zips through space, particle-wave, strikes the planet, angles on the continent, and filters through a mesh of land dust: clay bits, sod bits, tiny wind-borne insects, bacteria, shreds of wing and leg, gravel dust, grits of carbon, and dried cells of grass, bark, and leaves. Reddened, the light inclines into this valley over the green western mountains; it sifts between pine needles on northern slopes, and through all the mountain blackjack oak and haw, whose leaves are unclenching, one by one, and making an intricate, toothed, and lobed haze. The light crosses the valley, threads through the screen on my open kitchen window, and gilds the painted wall. A plank of brightness bends from the wall and extends over the goldfish bowl on the table where I sit. The goldfish's side catches the light and bats it my way; I've an eyeful of fish-scale and star.

This Ellery cost me twenty-five cents. He is a deep red-orange, darker than most goldfish. He steers short distances mainly with his slender red lateral fins; they seem to provide impetus for going backward, up, or down. It took me a few days to discover his ventral fins; they are completely transparent and all but invisible—dream fins. He also has a short anal fin, and a tail that is deeply notched and perfectly transparent at the two tapered tips. He can extend his mouth, so that it looks like a length of pipe; he can shift the angle of his eyes in his head so he can look before and behind himself, instead of simply out to his side. His belly, what there is of it, is white ventrally, and a patch of this white extends up his sides—the variegated Ellery. When he opens his gill slits he shows a thin crescent of silver where the flap overlapped—as though all his brightness were sunburn.

For this creature, as I said, I paid twenty-five cents. I had never bought an animal before. It was very simple; I went to a store in Roanoke called "Wet Pets"; I handed the man a quarter, and he handed me a knotted plastic bag bouncing with water in which a green plant floated and the goldfish swam. This fish, two bits' worth, has a coiled gut, a spine radiating fine bones, and a brain. Just before I sprinkle his food flakes into his bowl, I rap three times on the bowl's edge; now he is conditioned, and swims to the surface when I rap. And, he has a heart.

Once, years ago, I saw red blood cells whip, one by one, through the capillaries in a goldfish's transparent tail. The goldfish was etherized. Its head lay in a wad of wet cotton wool; its tail lay on a tray under a dissecting microscope, one of those wonderful light-gathering microscopes with two eyepieces like a stereoscope in which the world's fragments— even the skin on my finger—look brilliant with myriads of colored lights, and as deep as any alpine landscape. The red blood cells in the goldfish's tail streamed and coursed through narrow channels invisible save for glistening threads of thickness in the general translucency. They never wavered or slowed or ceased flowing, like the creek itself; they streamed redly around, up, and on, one by one, more, and more, without end. (The energy of that pulse reminds me of something about the human body: if you sit absolutely perfectly balanced on the end of your spine, with your legs either crossed tailor-fashion or drawn up together, and your arms forward on your legs, then even if you hold your breath, your body will rock with the energy of your heartbeat, forward and back, effortlessly, for as long as you want to remain balanced.) Those red blood cells are coursing in Ellery's tail now, too, in just that way, and through his mouth and eyes as well, and through mine. I've never forgotten the sight of those cells; I think of it when I see the fish in his bowl; I think of it lying in bed at night, imagining that if I concentrate enough I might be able to feel in my fingers' capillaries the small knockings and flow of those circular dots, like a string of beads drawn through my hand.

●

Margaret Mead

In her classic work, Coming of Age in Samoa, *anthropologist Margaret Mead used a primitive society as a laboratory to study the questions "Are the disturbances which vex our adolescents due to the nature of adolesence itself or to civilization?" and "Under different conditions does adolescence present a different picture?" In a passage from the chapter called "Our Educational Problems in the Light of Samoan Contrasts," we see her beginning to draw her conclusions and stumbling upon many controversial issues as she does so. Remember that the social and ethical issues in the work reflect the time of the study, 1926.*

For many chapters we have followed the lives of Samoan girls, watched them change from babies to baby-tenders, learn to make the oven and weave fine mats, forsake the life of the gang to become more active members of the household, defer marriage through as many years of casual love-making as possible, finally marry and settle down to rearing children who will repeat the same cycle. As far as our material permitted, an experiment has been conducted to discover what the process of development was like in a society very different from our own. Because the length of human life and the complexity of our society did not permit us to make our experiment here, to choose a group of baby girls and bring them to maturity under conditions created for the experiment, it was necessary to go instead to another country where history had set the stage for us. There we found girl children passing through the same process of physical development through which our girls go, cutting their first teeth and losing them, cutting their second teeth, growing tall and ungainly, reaching puberty with their first menstruation, gradually reaching physical maturity, and becoming ready to produce the next generation. It was possible to say: Here are the proper conditions for an experiment; the developing girl is a constant factor in America and in Samoa; the civilization of America and the civilization of Samoa are different. In the course of development, the process of growth by which the girl baby becomes a grown woman, are the sudden and conspicuous bodily changes which take place at puberty accompanied by a development which is spasmodic, emotionally charged, and accompanied by an awakened religious sense, a flowering of idealism, a great desire for assertion of self against authority— or not? Is adolescence a period of mental and emotional distress for the growing girl as inevitably as teething is a period of misery for the small baby? Can we think of adolescence as a time in the life history of every girl child which carries with it symptoms of conflict and stress as surely as it implies a change in the girl's body?

Following the Samoan girls through every aspect of their lives, we have tried to answer this question, and we found throughout that we had to answer it in the negative. The adolescent girl in Samoa differed from her sister who had not reached puberty in one chief respect, that in the older girl certain bodily changes were present which were absent in the younger girl. There were no other great differences to set off the group passing through adolescence from the group which would become adolescent in two years or the group which had become adolescent two years before.

And if one girl past puberty is undersized while her cousin is tall and able to do heavier work, there will be a difference between them, due to their different physical endowment, which will be far greater than that which is due to puberty. The tall, husky girls will be isolated from her companions, forced to do longer, more adult tasks, rendered shy by a change of clothing, while her cousin, slower to attain her growth, will still be treated as a child and will have to solve only the slightly fewer problems of childhood. The precedent of educators here who recommended special tactics in the treatment of adolescent girls translated into Samoan terms would read: Tall girls are different from short girls of the same age, we must adopt a different method of educating them.

But when we have answered the question we set out to answer we have not finished with the problem. A further question presents itself. If it is proved that adolescence is not necessarily a specially difficult period in a girl's life—and proved it is if we can find any society in which that is so—then what accounts for the presence of storm and stress in American adolescents? First, we may say quite simply, that there must be something in the two civilizations to account for the difference. If the same process takes a different form in two different environments, we cannot make any explanations in terms of the process, for that is the same in both cases. But the social environment is very different and it is to it that we must look for an explanation. What is there in Samoa which is absent in America, what is there in America which is absent in Samoa, which will account for this difference?

Such a question has enormous implications and any attempt to answer it will be subject to many possibilities of error. But if we narrow our question to the way in which aspects of Samoan life which irremediably affect the life of the adolescent girl differ from the forces which influence our growing girls, it is possible to try to answer it.

The background of these differences is a broad one, with two important components; one is due to characteristics which are Samoan, the other to characteristics which are primitive.

The Samoan background which makes growing up so easy, so simple a matter, is the general casualness of the whole society. For Samoa is a place where no one plays for very high stakes, no one pays very heavy prices, no one suffers for his convictions or fights to the death for special ends. Disagreements between parent and child are settled by the child's moving across the street, between a man and his village by the man's removal to the next village, between a husband and his wife's seducer by a few fine mats. Neither poverty nor great disasters threaten the people to make them hold their lives dearly and tremble for continued existence. No implacable gods, swift to anger and strong to punish, disturb the even tenor of their days. Wars and cannibalism are long since passed away and now the greatest cause for tears, short of death itself, is a journey of a relative to another island. No one is hurried along in life or punished harshly for slowness of development. Instead the gifted, the precocious, are held back, until the slowest among them have caught the pace. And in personal relations, caring is as slight. Love and hate, jealousy and revenge, sorrow and bereavement, are all matters of weeks. From the first months of its life, when the child is handed carelessly from one woman's hands to another's, the lesson is learned of not caring for one person greatly, not setting high hopes on any one relationship.

And just as we may feel that the Occident penalizes those unfortunates who are born into Western civilizations with a taste for meditation and a complete distaste for activity, so we may say that Samoa is kind to those who have learned the lesson of not caring, and hard upon those few individuals who have failed to learn it. Lola and Mala and little Siva, Lola's sister, all were girls with a capacity for emotion greater than their fellows. And Lola and Mala, passionately desiring affection and too violently venting upon the community their disappointment over their lack of it, were both delinquent, unhappy misfits in a society which gave all the rewards to those who took defeat lightly and turned to some other goal with a smile.

In this casual attitude towards life, in this avoidance of conflict, of poignant situations, Samoa contrasts strongly not only with America but also with most primitive civilizations. And however much we may deplore such an attitude and feel that important personalities and great art are not born in so shallow a society, we must recognize that here is a strong factor in the painless development from childhood to womanhood. For where no one feels very strongly, the adolescent will not be tortured by poignant situations. There are no such disastrous choices as those which confronted young people who felt that the service of God demanded forswearing the world forever, as in the Middle Ages, or cutting off one's finger as a religious offering, as among the Plains Indians. So, high up in our list of explanations we must place the lack of deep feeling which the Samoans have conventionalized until it is the very framework of all their attitudes toward life.

And next there is the most striking way in which all isolated primitive civilizations and many modern ones differ from our own, in the number of choices which are permitted to each individual. Our children grow up to find a world of choices dazzling their unaccustomed eyes. In religion they may be Catholics, Protestants, Christian Scientists, Spiritualists, Agnostics, Atheists, or even pay no attention at all to religion. This is an unthinkable situation in any primitive society not exposed to foreign influence. There is one set of gods, one accepted religious practice, and if a man does not believe, his only recourse is to believe less than his fellows; he may scoff but there is no new faith to which he may turn. Present-day Manu'a approximates this condition; all are Christians of the same sect. There is no conflict in matters of belief although there is a difference in practice between Church-members and non-Church-members. And it was remarked that in the case of several of the growing girls the need for choice between these two practices may some day produce a conflict. But at present the Church makes too slight a bid for young unmarried members to force the adolescent to make any decision.

Similarly, our children are faced with half a dozen standards of morality: a double sex standard for men and women, a single standard for men and women, and groups which advocate that the single standard should be freedom while others argue that the single standard should be absolute monogamy. Trial marriage, companionate marriage, contract marriage — all these possible solutions of a social impasse are paraded before the growing children while the actual conditions in their own communities and the moving pictures and magazines inform them of mass violations of every code, violations which march under no banners of social reform.

The Samoan child faces no such dilemma. Sex is a natural, pleasurable thing; the freedom with which it may be indulged in is limited by just one consideration, social status. Chiefs' daughters and chiefs' wives should indulge in no extra-marital experiments. Responsible adults, heads of households, and mothers of families should have too many important matters on hand to leave them much time for casual amorous adventures. Every one in the community agrees about the matter, the only dissenters are the missionaries who dissent so vainly that their protests are unimportant. But as soon as a sufficient sentiment gathers about the missionary attitude with its European standard of sex behavior, the need for choice, the forerunner of conflict, will enter into Samoan society.

•

174

Lewis Carroll

Among the many conundrums, riddles, illusions, and other verbal acrobatics employed by mathematician Charles Lutwidge Dodgson (alias Lewis Carroll, 1832-98) in Through the Looking Glass *is this passage, where cause and effect are reversed.*

"Living backwards!" Alice repeated in great astonishment. "I never heard of such a thing!"

"—but there's one great advantage in it, that one's memory works both ways."

"I'm sure *mine* only works one way," Alice remarked. "I can't remember things before they happen."

"It's a poor sort of memory that only works backwards," the Queen remarked.

"What sort of things do *you* remember best?" Alice ventured to ask.

"Oh, things that happened the week after next," the Queen replied in a careless tone. "For instance, now," she went on, sticking a large piece of plaster on her finger as she spoke, "there's the King's Messenger. He's in prison now, being punished: and the trial doesn't even begin till next Wednesday: and of course the crime comes last of all."

"Suppose he never commits the crime?" said Alice.

"That would be all the better, wouldn't it?" the Queen said, as she bound the plaster round her finger with a bit of ribbon.

Alice felt there was no denying *that.* "Of course it would be all the better," she said, "but it wouldn't be all the better his being punished."

"You're wrong *there*, at any rate," said the Queen. "Were *you* ever punished?"

"Only for faults," said Alice.

"And you were all the better for it, I know!" the Queen said triumphantly.

"Yes, but then I *had* done the things I was punished for," said Alice: "that makes all the difference."

"But if you *hadn't* done them," the Queen said, "that would have been better still; better, and better, and better!" Her voice went higher with each "better," till it got quite to a squeak at last.

Alice was just beginning to say "There's a mistake somewhere——," when the Queen began screaming, so loud that she had to leave the sentence unfinished. "Oh, oh, oh!" shouted the Queen, shaking her hand about as if she wanted to shake it off. "My finger's bleeding! Oh, oh, oh, oh!"

Her screams were so exactly like the whistle of a steam-engine, that Alice had to hold both her hands over her ears.

"What *is* the matter?" she said, as soon as there was a chance of making herself heard. "Have you pricked your finger?"

"I haven't pricked it *yet*," the Queen said, "but I soon shall — oh, oh, oh!"

"When do you expect to do it?" Alice asked, feeling very much inclined to laugh.

"When I fasten my shawl again," the poor Queen groaned out: "the brooch will come undone directly. Oh, oh!" As she said the words the brooch flew open, and the Queen clutched wildly at it, and tried to clasp it again.

"Take care!" cried Alice. "You're holding it all crooked!" And she caught at the brooch; but it was too late: the pin had slipped, and the Queen had pricked her finger.

"That accounts for the bleeding, you see," she said to Alice with a smile. "Now you understand the way things happen here."

"But why don't you scream *now*?" Alice asked, holding her hands ready to put over her ears again.

"Why, I've done all the screaming already," said the Queen. "What would be the good of having it all over again?"

•

Edwin Abbott

"Superstring" theory, a modern theory of physics, proposes that the fundamental stuff of nature consists of tiny strings that vibrate in six dimensions beyond our own, dimensions that we cannot perceive. (For elaboration, see the October 18, 1987 New York Times article "A Theory of Everything," by K.C. Cole.) Edward Witten, of the Institute for Advanced Study at Princeton, the main proponent of the theory, has been recorded as saying in his lectures, "We've been living in a finite dimensional world. Now I'm inviting you to jump into a world of infinite dimension." To his students, perhaps his ideas are like those of the sphere to the square in a novel published in 1884 called Flatland, a Romance of Many Dimensions. Written by a schoolmaster, Edwin

Abbott, whose primary interests were literature and theology, it brings the kind of energy to mathematics that Through the Looking Glass *brings from mathematics to literature. Here is a bit of dialogue from* Flatland.

I. My Lord, your assertion is easily put to the test. You say I have a Third Dimension, which you call "height." Now, Dimension implies direction and measurement. Do but measure my "height," or merely indicate to me the direction in which my "height" extends, and I will become your convert. Otherwise, your Lordship's own understanding must hold me excused.

Stranger. (To himself.) I can do neither. How shall I convince him? Surely a plain statement of facts followed by ocular demonstration ought to suffice. — Now, Sir; listen to me.

You are living on a Plane. What you style Flatland is the vast level surface of what I may call a fluid, on, or in, the top of which you and your countrymen move about, without rising above it or falling below it.

I am not a plane Figure, but a Solid. You call me a Circle; but in reality I am not a Circle, but an infinite number of Circles, of size varying from a Point to a Circle of thirteen inches in diameter, one placed on the top of the other. When I cut through your plane as I am now doing, I make in your plane a section which you, very rightly, call a Circle. For even a Sphere — which is my proper name in my own country — if he manifest himself at all to an inhabitant of Flatland — must needs manifest himself as a Circle.

Do you not remember — for I, who see all things, discerned last night the phantasmal vision of Lineland written upon your brain — do you not remember, I say, how, when you entered the realm of Lineland, you were compelled to manifest yourself to the King, not as a Square, but as a Line, because that Linear Realm had not Dimensions enough to represent the whole of you, but only a slice or section of you? In precisely the same way, your country of Two Dimensions is not spacious enough to represent me, a being of Three, but can only exhibit a slice or section of me, which is what you call a Circle.

The diminished brightness of your eye indicates incredulity. But now prepare to receive proof positive of the truth of my assertions. You cannot indeed see more than one of my sections, or Circles, at a time: for you have no power to raise your eye out of the plane of Flatland; but you can at least see that, as I rise in Space, so my sections become smaller. See now, I will rise; and the effect upon your eye will be that my Circle will become smaller and smaller till it dwindles to a point and finally vanishes.

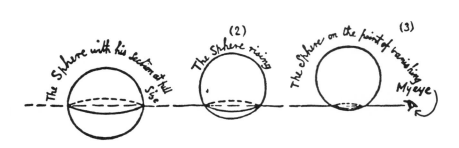

There was no "rising" that I could see; but he diminished and finally vanished. I winked once or twice to make sure that I was not dreaming. But it was no dream. For from the depths of nowhere came forth a hollow voice—close to my heart it seemed—"Am I quite gone? Are you convinced now? Well, now I will gradually return to Flatland and you shall see my section become larger and larger."

Every reader in Spaceland will easily understand that my mysterious Guest was speaking the language of truth and even of simplicity. But to me, proficient though I was in Flatland Mathematics, it was by no means a simple matter. The rough diagram given above will make it clear to any Spaceland child that the Sphere, ascending in the three positions indicated there, must needs have manifested himself to me, or to any Flatlander, as a Circle, at first of full size, then small, and at last very small indeed, approaching to a Point. But to me, although I saw the facts before me, the causes were as dark as ever. All that I could comprehend was, that the Circle had made himself smaller and vanished, and that he had now reappeared and was rapidly making himself larger.

When he regained his original size, he heaved a deep sigh; for he perceived by my silence that I had altogether failed to comprehend him. And indeed I was now inclining to the belief that he must be no Circle at all, but some extremely clever juggler; or else that the old wives' tales were true, and that after all there were such people as Enchanters and Magicians.

•

F. Woodbridge Constant

In the preface to Fundamental Principles of Physics *(1967) "a textbook for a one-semester liberal arts physics course at the college level," author F. Woodbridge Constant says, "It is indeed hoped that readers of this book will want to explore further the exciting world of physics and to follow its future development, and that they will find the study of physics interesting and enjoyable." He bears witness that such interest and enjoyment are feasible when he describes a theory proposed by Einstein to explain the photoelectric effect. The analogies he uses are similar to those employed by students in the Visualization exercise in the Essay Development Workshop section.*

In 1905 Einstein proposed another theory, one suggested by Planck's work on blackbody radiation. If, argued Einstein, radiation of frequency f is emitted in units of energy equal to hf and if it is absorbed by the oscillators of the cavity walls in units of the same energy, then why should we not regard the energy of an electromagnetic wave of frequency f as being concentrated in small bundles, each of energy hf? These bundles of energy are called *photons* or *light quanta*. According to this theory, an increase in the *intensity* of a beam of light implies an increase in the *number* of photons passing through a unit cross-sectional area of the beam in unit time, while an increase in the *frequency* of the light corresponds to an increase in the amount of *energy* carried *per photon*. As an analogy, consider two machine guns, A and B, and let B fire bullets of higher calibre than A. The bullets correspond to the photons in a beam of light. If either gun is caused to fire more bullets per second, the total energy output, or intensity of the firing, will increase, but since the energy of each bullet depends only on the powder in its case, the energy of the individual bullets will not be affected by the rate of firing. On the other hand, the bullets from gun B will possess more energy than those from gun A. Shifting from gun A to gun B corresponds in the case of light to increasing the frequency of the light.

Applying the photon theory to the photoelectric effect, Einstein reasoned as follows: If light of frequency f falls on the surface of a metal, then concentrated bundles of energy, each of magnitude hf, must strike the metal and penetrate to where they are stopped by the electrons in the metal. Thus a photon of energy hf may give all of its energy to one electron and the latter may use this energy to escape through the surface of the metal. The work function W represents the portion of the energy given to the electron that must be used in getting the electron out of the metal. What is left over should, according to the conservation of energy principle, be the KE of the emitted electron. The above reasoning may be expressed mathematically by the equation

$hf - W = $ KE of photoelectron,

called *Einstein's photoelectric equation;* W varies with the metal used, but not with the frequency of the light. The threshold frequency f_0 is that for which the electrons escape with negligible KE, or

$hf_0 = W.$

A good analogy to the photoelectric effect is that of a group of prisoners in jail, each under a bond W. A friend arrives, goes at random to one of the group, and offers to loan him x dollars. If x is less than W, the loan is declined; accepting the money will not enable the prisoner to get out of jail and he cannot spend money in jail. However, if x is greater than the bond W, the prisoner eagerly accepts the x dollars, pays his bond and leaves the jail with $(x - W)$ dollars in his pocket, which he may spend at the tavern of his choice!

•

R. Buckminster Fuller

R. Buckminster Fuller (b. 1895), an architect and inventor, is best known for his geodesic dome. But he is also a poet. The following excerpt from his long autobiographical poem, "How Little I Know," is typical in its informal language, modern syntax, line-breaks, and subject matter that is both complex and everyday.

And later I discovered that
Eddington had said "*Science* is:
The conscientious attempt
To set in order
The facts of *Experience*."

And I also discovered
That Ernst Mach—
The great Viennese physicist,
Whose name is used

To designate flight velocity
In *speed of sound* increments,
Known as Mach numbers—
Said:
"*Physics* is:
Experience
Arranged in
Most economical order."

So I realized that
Both Eddington and Mach
Were seeking to put in *order*
The same "raw materials"—
I.e. Experiences—
With which to identify
Their special subsystems
Of UNIVERSE.

Wherefore I realized that
All the words in all dictionaries
Are the consequent tools
Of all men's conscious
And conscientious attempts
To communicate
All their experiences—
Which is of course
To communicate
Universe.

There are forty-three thousand current words
In the Concise Oxford Dictionary.
We don't know who invented them!
What an enormous, anonymous inheritance!
Shakespeare used ten thousand of them
With which to formulate
His complete "works."
It would take many more volumes
Than Shakespeare's to employ
The forty-three thousand—
Logically and cogently.

•

An Annotated Science Writing Bibliography

This bibliography is not intended to be a complete list of all the classics of science writing. It is simply an annotated list of the books we have found to be inspiring and useful in introducing writing and writing experiments into the science, math, and technology curriculums in ways that have not been done consistently in most American schools.

Abbott, Edwin A. *Flatland: A Romance of Many Dimensions.* New York: Barnes & Noble, 1963. A first-person narrative about life in two dimensions and its expansion to the sphere. A classic fantasy about geometry.

Agassiz, Louis. *Essay on Classification.* Cambridge: Harvard Univ. Press, 1962. An essay on the system of arranging plants and animals into natural, related groups. Also, see his *Geological Sketches* and *Methods of Study in Natural History.*

Akimushkin, Igor. *Animal Travellers.* Moscow: Mir Publishers, 1970. About the method and reason in the migrations and wanderings of ancient and contemporary animals and insects. Excellent titles, beginnings and endings of individual essays. Contains astonishing information.

Amadon, Alfred M. *The Fold-Out Atlas of the Human Body.* New York: A Bonanza Pop-Up Book, Crown, 1984. Beautiful full-color replicas of the human body in the form of multi-layered, illustrated, movable fold-out plates. A reprint of the 1906 edition of the *Atlas of Physiology and Anatomy of the Human Body.*

Anderson, Edgar. *Landscape Papers*. Berkeley: Turtle Island Foundation, 1976. A series of short essays including much on the natural history of cities and the "acceptance of cities as places to live right in the middle of." The essay "The City Watcher" is a meditation on observing man in nature. Anderson is also the author of *Plants, Man, and Life*.

Audubon, John James. *Selected Birds and Quadrupeds of North America*. Montreal: Optimum Publishing Co., 1978. One of many interesting collections of this great naturalist-artist's work. A good way to see how it was done when sciences such as ornithology were in their infancy. See also his *Journals*.

Bakker, Robert. *The Dinosaur Heresies: New Theories Unlocking the Mystery of the Dinosaurs and Their Extinction*. New York: Morrow, 1986.

Bateson, Gregory. *Mind and Nature: A Necessary Unity*. New York: Dutton, 1979. Bateson asserts that biological evolution is a mental process. This book is worth reading for its range of ideas and disciplines that become interrelated, and for the clarity of its prose.

————. *Steps to an Ecology of Mind*. New York: Ballantine, 1972. A collection of essays about the nature of order and organization in living organisms and social systems: anthropology, psychiatry, biological evolution, genetics, and epistemology.

Bateson, Gregory and Mary Catherine Bateson. *Angels Fear*. New York: Macmillan, 1987. These essays extend the themes of Gregory Bateson's *Mind and Nature* to show how they apply to art, language, and religion. The sections on metaphor are particularly interesting.

Benecerraf, Paul and Hilary Putnam, eds. *Philosophy of Mathematics: Selected Readings*. Englewood, NJ: Prentice-Hall, 1964.

Bohren, Craig F. *Clouds in a Glass of Beer: Simple Experiments in Atmospheric Physics*. New York: Wiley, 1987. Answers to questions about natural phenomena.

Brand, Stewart. *The Media Lab: Inventing the Future at MIT*. New York: Viking, 1987. Conversations with machines and mad scientists.

Brillat-Savarin, Jean Anthelme. *The Physiology of Taste*. New York: Dover, 1960. The science and art of cooking and eating, subtitled *Meditations on Transcendental Gastronomy*. Written in the eighteenth century. Includes notes on the senses, on taste and thirst, on the end of the world, on the history of cooking. The section "on dreams" is a fine and hilarious example of anecdotal scientific style. This book has numbered sections, marginal notes, and a wealthy point of view.

Broad, C.D. *Lectures on Psychical Research.* London: Routledge & Kegan Paul.

Bronowski, J. and M. Selsam. *Biography of an Atom.* New York: Harper & Row, 1987. Grades 4–7.

Brown, A.E. & H.A. Jeffcott, Jr. *Absolutely Mad Inventions.* New York: Dover, 1970.

Brown, Tom. *The Tracker.* New York: Berkley, 1979. Traditional path-finding in the modern woods of New Jersey's Pine Barrens. By the author of *Tom Brown's Field Guide to City and Suburban Survival.*

Bryant, Mark. *Riddles Ancient and Modern.* New York: Peter Bedrick Books, 1983. An anatomy, history, and collection of riddles.

> The beginning of eternity,
> The end of time and space,
> The beginning of every end,
> And the end of every place.

(Answer: the letter "e")

Buckley, Walter, ed. *Modern Systems Research for the Behavioural Scientist.* Hawthorne, NY: Aldine, 1968. Essays on the development of an overall view of the relationships between human and animal behavior and the environment.

Byrd, Richard E. *Little America.* New York: Putnam's, 1930. A fascinating account of the planning and carrying out of the first modern aerial exploration of the South Pole. An exciting example of the interspersing of photographs, charts, lists, diaries, letters, and discursive prose.

————. *Alone.* New York: Putnam's, 1938. A diary of the scientific experiments and personal experiences of Byrd's months spent manning the advance weather base in the Antarctic winter night of 1934. This text can be used to show the value of diaries in scientific studies, as well as the metaphorical and real risks involved in such exploration and experimentation.

Capra, Fritjof. *Tao of Physics.* New York: Bantam, 1977.

Carroll, Lewis. *Through the Looking-Glass and What Alice Found There.* New York: Random House, 1946. The classic work by the mathematician Charles Lutwidge Dodgson. In effect, an excellent math and logic text.

Carson, Rachel. *The Sea Around Us.* New York: Science Library, 1961. Carson's classic science of the sea, beginning: "Beginnings are apt to be shadowy, and so it is with the beginnings of that great mother of life, the sea." Science, poetic language, perfect essay structure, illustration, and a useful appendix make this book a great example of popular science writing: "The sea is blue because the sunlight is reflected back to

184

our eyes from the water molecules or from very minute particles suspended in the sea. In the journey of the light rays into deep water all the red rays and most of the yellow rays of the spectrum have been absorbed, so when the light returns to our eyes it is chiefly the cool blue rays that we see." See also her *Silent Spring*.

Carter, Luther J. *Nuclear Imperatives and Public Trust: Dealing with Radioactive Waste*. Baltimore: Johns Hopkins, 1987.

Chomsky, Noam. *Language and the Mind*. New York: Harcourt Brace, 1968.

Coker, R.E. *This Great and Wide Sea*. New York: Harper & Row, 1947. An introduction to oceanography and marine biology, including a wonderful opening history of oceanographic pioneers. Interesting maps.

Constant, F. Woodbridge. *Fundamental Principles of Physics*. Reading: Addison-Wesley, 1967. A physics for poets textbook, simply written, using good analogies, but dated now.

Cornell, James and John Surowiecki, eds. *The Pulse of the Planet: A State of the Earth Report from the Smithsonian Institution Center for Short-Lived Phenomena*. New York: Crown Harmony, 1972. A state-of-the-planet report on volcanoes, earthquakes, starfish plagues, oil spills, tidal waves, meteorite falls, fireballs, moth infestations, species' extinctions, landslides, forest fires, red tides, animal migrations, floods and storm surges and other urgent and unusual events relating to the ecosystem.

Curtis, Brian. *The Life Story of the Fish: His Manners and Morals*. New York: Dover, 1949.

Darwin, Charles. *The Origin of Species by Means of Natural Selection or the Preservation of Favoured Races in the Struggle for Life*. London: J. Murray, 1859. That's the original. Many reprints are available, including a good one by Avenel Books, which includes a historical sketch by Darwin.

———. *The Various Contrivances by Which Orchids Are Fertilized by Insects*. Chicago: Univ. of Chicago Press, 1984.

Darwin, Charles and T.H. Huxley. *Autobiographies*. New York: Oxford Univ. Press, 1983. A beautiful contrast in writing styles.

Da Vinci, Leonardo. *The Notebooks of Leonardo Da Vinci*. Jean Paul Richter, ed. 2 vols. New York: Dover, 1970. The original notebooks in English and Italian profusely illustrated by Leonardo. Includes sections on anatomy, zoology, physiology, astronomy, physical geography, naval warfare, flying machines, plus humorous writings, lies, and Leonardo's shopping lists. A useful text for the illustration of the process of combining science writing with visual examples. Also a great example of the attempt at the perception of the whole by a single scientist.

DeBono, Edward. *New Think*. New York: Basic Books, 1967. The uses of "lateral" thinking in generating new ideas. Breaking out of restrictive and habitual thought patterns.

Dillard, Annie. *Pilgrim at Tinker Creek*. New York: Harper & Row, 1974. A book of concrete observations and mystical insights by a poetic writer. Naturalism in the tradition of *Walden Pond*.

———. *Teaching a Stone to Talk: Expeditions and Encounters*. New York: Harper and Row, 1982. The chapter "An Expedition to the Pole" can be useful in studying and teaching the structure of the essay.

Dodd, Robert T. *Thunderstones and Shooting Stars*. Cambridge: Harvard Univ. Press, 1986. The nature and origin of meteorites.

Ehrlich, Paul and Anne Ehrlich. *Extinction: The Causes and Consequences of the Disappearance of Species*. New York: Ballantine, 1983. A concise, thorough, and lucid study of the loss of species, which are disappearing at a rate unprecedented in history. Proposes what can be done about it.

Einstein, Albert. "Autobiographical Notes" in *Albert Einstein: Philosopher-Scientist*. Evanston: Library of Living Philosophers, Vol. VII, 1949. A history of the development of Einstein's thought, including his delineations of what is thinking and what is wonder. With facing German text.

———. *Ideas and Opinions*. New York: Dell, 1954. An anthology of Einstein's general writings on minorities, scientific truth, atoms, space, freedom, education, pacifism, good and evil, and so on. Includes his 1939 World's Fair time capsule message. In this book, Einstein's writing style is a mixture of practicality and spirit.

Eiseley, Loren. *The Immense Journey*. New York: Random House, 1957. Also, see his *Notes of an Alchemist, Unexpected Universe*, and *The Man Who Saw Through Time*.

Eliade, Mircea. *Myths, Dreams, and Mysteries*. London: Fontana, 1960. The relation between ancient and modern man's experience of the mysterious.

Emerson, Ralph Waldo. *Nature*. Boston: Beacon Press, 1985. A facsimile edition of Emerson's first book. "Every natural action is graceful" and "Man is the dwarf of himself."

Fabré, J. Henri. *The Sacred Beetle and Others*. New York: Dodd, Mead and Co., 1918.

Feinberg, Gerald. *The Prometheus Project*. New York: Anchor, 1969. An examination of humanity's short- and long-range goals, including discussions of agreement and the unity of mankind.

Feynman, Richard P. *"Surely You're Joking, Mr. Feynman!"* New York: Norton, 1985. Extremely entertaining personal essays by a Nobel scientist whose down-to-earth intellectual curiosity is inspiring. "I learned to pick locks from a guy named Leo Lavatelli."

Fermi, Laura. *Atoms in the Family.* Chicago: Univ. of Chicago, 1954.

Foucault, Michel. *The Order of Things: An Archaeology of the Human Sciences.* New York: Vintage, 1970. Systems classification.

Freud, Sigmund. *The Interpretation of Dreams.* New York: Avon, 1965. The first systematic attempt to make a science of the analysis of dreams. Freud uses himself as the subject. A one-of-a-kind scientific text in which the poetry of dreaming is combined with first-rate science writing. The reading of the table of contents alone is inspiring.

———. *Leonardo Da Vinci.* New York: Vintage, 1969. Freud's amazing psychosexual analysis of the Mona Lisa, and the vulture fantasy. An exciting introduction to Freud's way of thinking, especially if you like art and scary books.

———. *An Outline of Psychoanalysis.* New York: Norton, 1949. This book describes the beginnings of psychoanalysis and its practical and theoretical implications. Beautifully and intensely written.

Fuller, John G. *Tornado Watch #211.* New York: Morrow, 1987. The scientific and journalistic recounting of a series of tornadoes on one day in 1985 in North America.

Fuller, R. Buckminster. *Untitled Epic Poem on the History of Industrialization.* Charlotte, NC: The Nantahala Foundation, 1962. A poem written in 1940, addressed to poets and scientists. This poem can be used to show how scientific information can become the subject of a modern poem, and how that information might be the only possible subject for a modern epic.

———. *And It Came to Pass—Not to Stay.* New York: Macmillan, 1976. Six long science poems by Fuller including "How Little I Know," "What I Am Trying to Do," and "A Definition of Evolution." This book provides an introduction to epistemology and science poetry.

Fulwiler, Toby, ed. *The Journal Book.* Portsmouth, NH: Boynton/Cook-Heinemann, 1987. The uses of journals in all kinds of classrooms, with a section on "Journals and the Quantitative Disciplines." Many student examples.

Gamow, George. *Thirty Years That Shook Physics.* New York: Doubleday Anchor, 1966. The story of the development of quantum theory told with clarity and humor. His *1,2,3...Infinity* is another good popularized book about science and math.

Gardner, Steven D. *The Urban Naturalist*. New York: Wiley, 1987. Plants and animals of American cities.

Gardner, Martin, ed. *The Sacred Beetle and Other Great Essays in Science*. Buffalo: Prometheus Books, 1984. A revised edition of *Great Essays in Science*, New York: Pocket Books, 1957. A feast of science writing, including some of the most perfect essays. This book alone could provide an entire year's curriculum in science writing. Includes work by Einstein, Lewis Thomas, Freud, Laura Fermi, Maeterlinck, H.G. Wells, Ortega y Gasset, Asimov, Oppenheimer, Whitehead, the Huxleys, Rachel Carson, William James, Stephen Jay Gould, Fabre, Krutch, Sagan, and others.

————. *Mathematical Puzzles and Diversions*. New York: Simon & Schuster, 1961.

————. *Martin Gardner's New Mathematical Diversions from Scientific American*. Chicago: Univ. of Chicago Press, 1984.

————. *Mathematical Carnival*. New York: Knopf, 1975.

————. *Mathematical Circus*. New York: Random House, 1981.

Gatty, Harold. *Nature Is Your Guide: How to Find Your Way on Land and Sea*. New York: Penguin, 1979. Written by the first man to fly around the world in eight days, this text, full of charts and illustrations, discusses the navigational techniques of primitive and modern people, including the uses of sound and smell, getting directions from trees, plants, anthills, and animals, estimation of distance and direction from hills, rivers, waves, and swells, plus clues from the shifting of sands, the patterns of snow fields and astronomical data. Sections on how to find your way in the desert, polar regions, and the city.

Giedion, Sigfried. *Mechanization Takes Command*. Oxford: Oxford Univ. Press, 1948 and New York: Norton, 1969. A contribution to the political history of the machine.

Glass, Bentley. *Progress or Catastrophe: The Nature of Biological Science and Its Impact on Human Society*. New York: Praeger, 1985. A fascinating collection of essays on such topics as the relation of biology to the physical sciences, indeterminacy in biology, and the role of biology in the Nuclear Age. Clearly and convincingly written. Other works by Glass include *Science & Ethical Values*, *The Timely and the Timeless*, and *Forerunners of Darwin*.

Gleick, James. *Chaos*. New York: Viking Penguin, 1987. A study of wildness, disorder, and irregularity in the science of the everyday world in an attempt to lead to a unified theory or "theory of everything," with emphasis on the work of Mitchell Feigenbaum. Lucid writing with lots of good anecdotes and quotations.

Goodall, Jane. *In the Shadow of Man*. Boston: Houghton Mifflin, 1971. A famous study of chimpanzees in Tanzania, characterized by courageous field work and close observation.

Goodman, Paul and Percival Goodman. *Communitas*. New York: Vintage, 1960. On the art and science of building cities. Grassroots illustrations and tables. Much data about New York City.

Gould, Stephen Jay. *Hen's Teeth and Horse's Toes: Further Reflections in Natural History*. New York: Norton, 1983. Or any other book by Gould. This one includes the full essay excerpted as an example in our Samples section: "Nonmoral Nature."

————. *An Urchin in the Storm*. New York: Norton, 1987. A series of essays about books and science by this teacher of biology, geology, and the history of science. The book includes portraits of four eminent biologists. Gould's other books include *Ever Since Darwin, The Panda's Thumb*, and *The Mismeasure of Man*.

Government Printing Office. *NASA-SP-171: Earth Photographs from Gemini XII* and *NASA-SP-250: This Island Earth*. (Government Printing Office, Washington D.C.) These books are inexpensive and filled with color plates of the earth and their interpretations.

Gray, Henry. *Gray's Anatomy*. New York: Crown, 1977. The premier text in human anatomy, an international bestseller for 100 years. 780 illustrations.

Gregory, R.L. *The Intelligent Eye*. New York: McGraw Hill, 1971. On the nature of seeing, with 3-D illustrations.

Grieve, Mrs. M. *A Modern Herbal*. 2 vols. New York: Dover, 1971. "The medicinal, culinary, cosmetic and economic properties, cultivation and folklore of herbs, grasses, fungi, shrubs and trees, with all their modern scientific uses." Facts, recipes, stories, illustrations, and etymologies. First published in 1931.

Grossinger, Richard. *The Night Sky*. San Francisco: Sierra Club Books, 1981. The science and anthropology of the stars and planets, including the history of occult, ancient, and western astronomy; science fiction; flying saucers; and the "pop star" cult.

Hanson, Norwood Russell. *Patterns of Discovery*. Cambridge: Cambridge Univ. Press, 1972. A study of the work habits of scientists.

Harrison, Edward. *Darkness at Night: A Riddle of the Universe*. Cambridge: Harvard Univ. Press, 1987. Why is the sky dark at night? Is the universe finite or infinite?

Hawking, Stephen W. *A Brief History of Time: From the Big Bang to Black Holes*. New York: Bantam, 1988. A clear introductory history of the investigation of time and infinity by one of today's most noted astophysicists. Contains only one equation.

Heath-Stubbs, John, and Philips Salman, eds. *Poems of Science*. New York: Viking Penguin, 1984. An anthology of science poems, including Keats' "There was an awful rainbow once in heaven . . . ," Edgar Allan Poe's "To Science," Whitman's "When I heard the learn'd astronomer," Gerald Manley Hopkins' "I am like a slip of comet . . . ," Marianne Moore's "Four Quartz Crystal Clocks," Auden's "Unpredictable but Providential," Chaucer's "Sound Waves" from *The House of Fame*, and so on.

Heidegger, Martin. *Poetry, Language, Thought*. New York: Harper & Row, 1971. The uncovering of existence, song as being, and the nature of "Thing-in-itself" uniting earth, sky, divinities, and mortals. The structure of the chapter "Building Dwelling Thinking" is especially notable. Many questions.

———. *What is Called Thinking?* New York: Harper & Row, 1968. Twenty-one lectures from the fifties, beautifully incorporating summaries, transitions, and etymologies.

Heisenberg, Werner. *Physics and Beyond: Encounters and Conversations*. New York: Harper & Row, 1971. An account of the development of quantum mechanics. Chapter 16 is especially recommended: "The Responsibility of the Scientist (1945-1950)." The whole is full of astonishing quotations of Heisenberg and his fellow scientists.

Henn, T.R., *The Apple and the Spectroscope*. New York: Norton, 1966. Lectures on poetry designed for students of science. Analyses of works by Donne, Marvell, Blake, Yeats, Shakespeare, and Pope, and passages from the Bible. Some interesting equations and diagrams, one of which relates girls to anodes and roses to cathodes. Somebody should write a book like this about modern poems.

Herbert, Frank. *Dune*. New York: Berkley Books, 1965. The original ecologically conscious classic sci-fi work.

Hey, A. J. and P. Walters. *The Quantum Universe*. Cambridge: Cambridge Univ. Press, 1986. About how nobody understands quantum mechanics.

Hoagland, Edward. *Notes from the Century Before*. New York: Ballantine, 1969. Nature writer Hoagland's diaries of a journey from New York City to the wilds of British Columbia, and back.

Hofstadter, Douglas. *Godel, Escher, Bach: An Eternal Golden Braid*. New York: Basic Books, 1979.

Holton, Gerald. *The Advancement of Science, and Its Burdens*. Cambridge: Cambridge Univ. Press, 1963. A series of essays on the history and philosophy of science.

Hunter, J.A. and Joseph S. Madachy. *Mathematical Diversions*. New York: Dover, 1975.

Huxley, Aldous. *The Art of Seeing.* Berkeley: Creative Arts, 1982.

Jonson, Ben. *The Alchemist.* New York: Norton, 1976.

Keller, Dolores Elaine. *Sex and the Single Cell.* New York: Bobbs Merrill, 1972. An introduction to the biology of sex. Frequent and excellent illustrations and a practical bibliography.

Kohl, Judith and Herbert Kohl. *The View from the Oak.* New York: Sierra Club, Scribner's, 1977. Inspired by Jacob von Uexkull's essay "A Stroll through the World of Animals and Men: A Picture Book of Invisible Worlds," this book gives the view, scientifically, from the stance of the Other. Great illustrations. How to perceive time as animals might. A world view free of the egocentric.

Kohl, Herbert. *Mathematical Puzzlements: Play and Invention with Mathematics.* New York: Schocken, 1987. More than 250 games, puzzles, paradoxes, and experiments divided into Patterns on the Plane; Number Patterns; Knots, Maps, and Connections; and Logic and Strategy. A book full of new ideas.

Kropotkin, Petr. *Mutual Aid.* Boston: Extending Horizons Books, no date given. Written around 1925, this book is a description of the concept of mutual aid among animals, insects, barbarians, savages, medieval and modern people. Full of zoological detail, written by a revolutionary Russian prince.

Levi, Primo. *The Periodic Table.* New York: Schocken, 1984. A beautiful series of narrational meditations on twenty-one of the elements. First person writing.

Lilly, John C. *The Center of the Cyclone.* New York: Julian Press, 1972. An exploratory memoir on induced altered states of consciousness in relation to the concept of the ego. See also his *Communication between Man and Dolphin: The Possibility of Talking with Other Species, The Mind of the Dolphin,* and *Metaprogramming in the Human Biocomputer.*

———. *The Scientist.* New York: Bantam, 1978. An autobiography by the famous experimenter with dolphin communication, neurophysiology, consciousness-expanding drugs, isolation tanks, and communication with extraterrestrial beings. Good writing examples are the chapters "Life Refuses Closure" and "Simulation of the Future of Man, Dolphin, and Whale."

Littleton Industries. *Physicians' Desk Reference.* Oradell, N.J.: Medical Economics Books, updated each year. The reference book doctors use when they prescribe medication. Includes information about every prescription drug and colorful illustrations. Good for investigators.

Lopez, Barry. *Arctic Dreams: Imagination and Desire in a Northern Landscape*. New York: Scribner's, 1986. The meditations of a sensitive, concerned naturalist who writes with modern scientific accuracy and nineteenth-century eloquence about every aspect of a poorly understood part of the globe.

Lorentz, H.A., Albert Einstein, H. Minkowski & H. Weyk. *Einstein: The Principle of Relativity*. New York: Dover, 1952. Original papers on the general and special theories. Einstein's essay "Cosmological Considerations on the General Theory of Relativity" can be used as an example of how to intersperse prose with mathematics.

Lucretius. *On the Nature of Things*. New York: Norton, 1977. An ontological poem written in Rome c. 60 B.C. on the origin of the world, fear of death, the development of society, and the philosophy of Epicurus.

Macrorie, Ken. *Twenty Teachers*. New York: Oxford Univ. Press, 1984. The stories of twenty teachers—including science, math, and writing teachers — who enabled their learners to do good works. Includes many writing examples.

Madachy, Joseph S. *Mathematics on Vacation*. New York: Scribner's, 1966.

Maeterlinck, Maurice. *Vie des Abeilles*. New York: French and European, no date given. In French.

Maimon, Elaine. *Writing and Learning in the Arts and Sciences*. Boston: Little, Brown, 1981.

Manning, Kenneth R. *Black Apollo of Science: The Life of Ernest Everett Just*. New York: Oxford Univ. Press, 1983. An excellent biography of a prominent marine biologist who was forced to overcome nearly insuperable obstacles of racism.

Martin, Nancy, et al. *Writing and Learning across the Curriculum*. Upper Montclair, N.J.: Boynton/Cook, 1979.

Marx, Karl. *Das Kapital*. 3 vols. New York: International Publishing Co., 1967. Especially "The Development of the Factory." The only unabridged English edition.

Matthiessen, Peter. *Wildlife in America*. New York: Viking, 1959. A historical survey of the white man's effect on Northern American wildlife. Good descriptions of past variety and abundance, documented with quotations and anecdotes, full of line drawings and lively science writing.

McPhee, John. *Oranges*. New York: Farrar, Straus & Giroux, 1966. The history of the botany and industry of oranges and orange juice. A good example of how to write about an ordinary subject with poetic spirit and memorable scientific implication.

———. *The Curve of Binding Energy.* New York: Ballantine, 1976. The work of Theodore B. Taylor.

———. *The Deltoid Pumpkin Seed.* New York: Ballantine, 1973. A true but seeming-to-be-fiction story about the conception and building of an aircraft that could have changed the world, now hidden in a secret New Jersey testing ground. One more squashed experiment, fascinating to read about.

———. *The Pine Barrens.* New York: Ballantine, 1967. A descriptive ecology of the idiosyncratic rural community of people descended from the Prussians in south central New Jersey.

Mead, Margaret. *Coming of Age in Samoa.* New York: Morrow Quill Paperbacks. 1973. A groundbreaking anthropological work focusing on adolescent girls in Samoa. It answers the questions, "Are the disturbances which vex adolescents due to the nature of adolescence itself or to the civilization? Under different conditions does adolescence present a different picture?"

———. *Male and Female: A Study of the Sexes in a Changing World.* New York: Dell, 1968. The sharing of an anthropologist's knowledge of the roles of the sexes in various cultures. The first introductory essay, "The Significance of the Questions We Ask," could be used as a model for speculative thought on these issues. "Each Family in a Home of Its Own" provides endless material for contemporary discussion. Author of numerous other books, including *Cooperation and Competition among Primitive People.*

Monroe, Robert A. *Journeys Out of the Body.* New York: Doubleday, 1971. An American businessman writes a description of his unexpected experiences with astral projection, in a combination of discursive prose and journals. A terrific introduction to how to write about personal experiences that relate to scientific exploration and experimentation.

Mowat, Farley. *People of the Deer.* New York: Bantam, 1981. And many other works, including *Never Cry Wolf,* which was made into a movie.

Mumford, Lewis. *The Myth of the Machine.* 2 vols. New York: Harcourt, Brace, 1966-1974.

Murchie, Guy. *Music of the Spheres.* New York: Dover, no date. The material universe from atom to quasar, simply explained, for grades 7-12.

National Geographic: *Atlas of North America: Space Age Portrait of a Continent.* Washington, D.C.: National Geographic, 1985. An atlas combining traditional political and regional topographical maps with pictures derived from the "remote sensing" technologies of weather satellites, the Apollo, Skylab and shuttle spacecraft, aircraft, and

Landsats, a series of five satellites orbiting as high as 570 miles, whose "eyes" make use of electromagnetic energy in the infrared portion of the spectrum, reflected as light and transmitted as heat. The Landsat pictures create a new way of perceiving the earth, beyond the sight even of an astronaut.

Orr, Eleanor Wilson. *Twice As Less*. New York: Norton, 1987. The relationship between black English and the performance of black students in mathematics and science.

Padgett, Ron, ed. *Handbook of Poetic Forms*. New York: Teachers & Writers Collaborative, 1987. A book of forms, ancient through modern, created for secondary school students and their teachers.

Panati, Charles. *Extraordinary Origins of Everyday Things*. New York: Harper & Row, 1987. Stories of the origins of customs, superstitions, kitchen and bathroom implements, central heating, seeds, cosmetics, marbles, buttons, top hats, graham crackers, and so on.

Papert, Seymour. *Mind Storms: Children, Computers, and Powerful Ideas*. New York: Basic Books, 1980. An in-depth examination of children's use of the LOGO computer language. Concentrates on math and science, but there is some discussion of writing prose and poetry.

Perry, Donald. *Life above the Jungle Floor*. New York: Simon & Schuster, 1986. Climbing trees in Costa Rica to study biology.

The Peterson Field Guide Series. Boston: Houghton Mifflin, various dates. Field Guides to birds, trees, shrubs, rocks and minerals, insects, fish, mammals, wildflowers, reptiles, amphibians, stars and planets, edible wild plants, shells, butterflies, animal tracks, and ferns. Designed for quick identification. Perhaps the most interesting guide for people in a city is *A Field Guide to the Atmosphere* that includes information and illustrations about cloud identification, snow and snow crystals, storms, sky colors, sunsets and sunrises, weather modification and phenomena, and experiments to perform.

Peterson, Roger Tory, and James Fisher. *Wild America*. Boston: Houghton Mifflin, 1955. The record of a 30,000-mile journey around the continent in 100 days of 1953 written in the form of a collaboration between two naturalists.

Porter, Eliot. *In Wildness Is the Preservation of the World*. New York: Sierra Club, Ballantine, 1962. Selected passages from Thoreau set alongside Eliot Porter's photographs of trees, leaves, plants, animals and water. Porter has published many other books, including *Antarctica*, a combination of diaries and photography.

Pribram, Karl H. *Languages of the Brain: Experimental Paradoxes and Principles in Neuropsychology*. Englewood, NJ: Prentice-Hall, 1971.

Rifkin, Jeremy. *Entropy*. New York: Bantam, 1980. Old and new world views on the concept of entropy. Especially good writing on nuclear fission and fusion, and a good essay on recycling.

Riordan, Michael. *The Hunting of the Quark*. New York: Simon & Schuster, 1987. A history of the discovery and acceptance of the quark as a particle.

Robbe-Grillet, Alain. *For a New Novel: Essays on Fiction*. New York: Grove Press, 1964.

Rucker, Rudy, ed. *Mathenauts: Tales of Mathematical Wonder*. New York: Arbor House, 1987. 23 mind-blowing short stories written to torture and delight all mathematics students and those who love the pleasures of puzzles. Includes work by Isaac Asimov, Greg Bear, Ian Watson, Martin Gardner, Frederik Pohl, and others. Rucker is the author of *Spacetime Donuts, The Sex Sphere, Master of Space and Time*, and ten other books.

Russell, Bertrand. *The ABC of Relativity*. 3rd ed. London: Allen & Unwin, 1969. An explanation of relativity for the layman. Also see his *Principles of Mathematics* and *Definition of Number*.

Sagan, Carl. *Cosmos*. New York: Ballantine Books, 1980. The best seller based on the TV series. Chock full of history, wonderments, speculations about everything mathematical and scientific.

———. *The Dragons of Eden: Speculations on the Evolution of Human Intelligence*. New York: Random House, 1977. Brains, memories, and cosmic communication. "The Future Evolution of the Brain" is a great example of a various essay.

Sauer, Carl O. *Man in Nature*. Berkeley: Turtle Island Foundation, 1975. An unusual geography textbook about North and Central America before the advent of the white man. Though intended for the middle grades, it's useful to everyone.

———. *Northern Mists*. Berkeley: Turtle Island Foundation, 1968. A beautifully detailed history of European seafaring explorations during the Middle Ages.

Schuh, Fred. *The Master Book of Mathematical Recreations*. New York: Dover, 1968.

Scientific American. *Cosmology + 1*. New York: Scientific American, 1977. A series of essays. "The Curvature of Space in a Finite Universe" is most beautifully structured and illustrated.

Sears, Peter. *Secret Writing: Keys to the Mysteries of Reading and Writing*. New York: Teachers & Writers Collaborative, 1986. About secret codes, ciphers, hieroglyphics, ancient languages, comics, and communicating with extraterrestrial beings. Written as a series of exercises for high school students.

Shortridge, Virginia, ed. *Songs of Science*. Boston: Marshall Jones Co., 1930. A precious anthology of science poems, divided into sections on philosophy, healing, scientific achievements (including Lawrence Lee's "The Subway Builders"), flying, cities, the relationship between science and art, the technology of war, and ending with a section on peace (which includes Robert Herrick's "The White Island"). Poems by William Blake, Walt Whitman, Euripides, Milton, Emerson, Robert Frost, Babette Deutsch, Siegfried Sassoon, Harriet Monroe, and many other poets.

Skeat, Rev. Walter W. *A Concise Etymological Dictionary of the English Language*. New York: Capricorn Books, 1963. Histories of the meanings of words, including distributions of words according to the languages from which they are derived. The dictionary also exists in larger editions.

Smullyan, Raymond. *What Is the Name of This Book?: The Riddle of Dracula and Other Logical Puzzles*. Englewood, NJ: Prentice-Hall, 1978. By the author of *The Chess Mysteries of Sherlock Holmes* (New York: Knopf, 1979).

Sobel, Michael I. *Light*. Chicago: Univ. of Chicago Press, 1987. A history of light.

Spamer, Claribel. *Easy Science Plays for Grade School*. Boston: Baker's Plays, 1960. Short, charmingly inane plays written to teach children about science. These plays could serve as models for more imaginative ones that students could write. Anyway, who could resist a book written by someone named Claribel Spamer?

Spencer-Brown, G. *Laws of Form*. New York: Dutton, 1979. Divagations on Boolean algebra, relating somewhat to computer uses.

Steiner, Rudolf. *Health and Illness*, vol. 2 *of Lectures to the Workmen*. Spring Valley, N.Y.: Anthroposophic Press, 1983. Steiner's holistic theories about science, education, and medicine in relation to the health hazards of nicotine, alcohol, and cocaine.

Stevens, Peter S. *Handbook of Regular Patterns: An Introduction to Symmetry in Two Dimensions*. Cambridge: MIT Press, 1981.

Stevenson, Robert Louis. *Dr. Jekyll and Mr. Hyde*. New York: Bantam Books, 1985. The classic, originally published in 1886. Resonates nicely against current misapplications of science and abuses of chemical substances.

Tart, Charles, ed. *Altered States of Consciousness*. New York: Wiley, 1969. An anthology of scientific essays on dream consciousness, meditation, waking and pre-sleep imagery, hypnosis, psychedelic drugs, the physiology of trance states. Includes Kilton Stewart's famous essay on problem solving through dreaming among the Senoi tribe of Malaya.

Taylor, John G. *Black Holes: The End of the Universe?* New York: Random House, 1973. A book full of questions.

Ternes, Alan, ed. *Ants, Indians, and Little Dinosaurs.* New York: Scribner's, 1975. An anthology of writing from *Natural History* magazine, including delights by Margaret Mead, N. Scott Momaday, and Oliver La Farge. An unusual and wonderful book. Divided into four parts: On Fauna and a Stone Age Friend; On Fossils, Famous and Obscure; On Natives and Naivete; On Famine, Floods, and the Future.

Thoreau, Henry David. *Walden.* New York: Macmillan, 1962. Many other editions of this classic by the different-drummer naturalist.

Tobias, Sheila. *Succeed with Math: Every Student's Guide to Conquering Math Anxiety.* New York: College Board Publications, 1987. The author, co-founder of the Innovative Math Anxiety Clinic and author of the previous *Overcoming Math Anxiety,* offers in her new book a broad range of methods for getting comfortable with math.

Toulmin, Stephen. *Physical Reality.* New York: Harper & Row, 1970. Philosophical essays on twentieth-century physics.

Vare, Ethlie Ann and Greg Ptacek. *Mothers of Invention.* New York: Morrow, 1988. A history of women inventors.

Verne, Jules. *Master of the World.* Mahwah, NJ: Watermill Press, 1985. This is a useful book if anybody wants to know how to become the master of the world. Includes good delineations of the impenetrable mysteries of science mixed with politics.

Von Frisch, Karl. *Bees: Their Vision, Chemical Senses, and Language.* Ithaca: Cornell Univ. Press, 1971.

———. *Dance Language and Orientation of Bees.* Cambridge: Harvard Univ. Press, 1967.

———. *The Dancing Bees: An Account of the Life and Senses of the Honey Bee.* New York: Harcourt, Brace, 1961.

Warner, William W. *Beautiful Swimmers.* New York: Viking Penguin, 1977. A natural history of the Atlantic blue crab, Chesapeake Bay, and the Bay's commercial watermen.

Watson, James D. *The Double Helix.* New York: Signet, 1968. A wonderfully gossipy account of the discovery of the structure of DNA by some wild young geniuses.

The Way Things Work: An Illustrated Encyclopedia of Technology. New York: Simon & Schuster, 1967. Explanations of the most basic and the most complex scientific and industrial processes that affect daily life, including Polaroid-Land Colorfilm, the fountain pen, quartz and atomic clocks, the automatic rifle, the washing machine, hosiery, gas and water meters, dry cleaning, jet engines, and radioactivity. Plus why does a ship float? Many illustrations. From the 1963 original German language edition but never out of date.

Weekley, Ernest. *An Etymological Dictionary of Modern English*. New York: Dover, 1967. 2 vols.

White, Gilbert. *Journals of Gilbert White*, ed. by Walter Johnson. Cambridge: M.I.T., 1970. The journals of this 18th-century English naturalist that led to his book *Natural History of Selborne*, published in 1789. Full of precision, brevity, and clarity.

Whitehead, Alfred North. *Modes of Thought*. New York: Macmillan, 1966. "Apart from detail, and apart from system, a philosophic outlook is the very foundation of thought and of life. The sort of ideas we attend to, and the sort of ideas which we push into the negligible background, govern our hopes, our fears, our control of behaviour. As we think, we live. This is why the assemblage of philosophic ideas is more than a specialist study. It moulds our type of civilization."

Whorf, B.L. *Language, Thought, and Reality*. Cambridge: M.I.T., 1956. The classic presentation of the Whorfian hypothesis that "the limits of my language are the limits of my world," with examples from Amerindian languages.

Wilford, John Noble. *The Riddle of the Dinosaur*. New York: Random House, 1987.

Witchell, Nicholas. *The Loch Ness Story*. New York: Viking Penguin, 1974. Scientific looks at a current phenomenon, including extensive quotations from journals and transcriptions, to illuminate their uses in a text.

Zaslavsky, Claudia. *Africa Counts: Number and Pattern in African Culture*. Westport, CT: Lawrence Hill, 1979.

APPENDIX

Using Writing in Mathematics
By Russel W. Kenyon

Writing has a place in mathematics teaching and learning. Having students write about a problem requires them to clarify their thoughts about how they will approach the problem. This writing makes the concepts in the problem clearer and sharper. The writing process then becomes an integral part of the thought process: the students begin to gather, formulate, and organize old and new knowledge, concepts, and strategies, and to synthesize this information as a new structure that becomes a part of their own knowledge network.

To promote this thought processing, math teachers should design activities to assure that writing becomes a normal part of the daily routine. The students will begin to gain the necessary writing skills as the writing takes place. They will also begin to expect writing to be a standard part of the mathematics program. The following ideas should help the mathematics teacher make writing an integral part of that program. The actual cognitive learning may be somewhat difficult to observe and measure. However, the teacher should begin to see increased student performance on word problem solving and in other critical thinking exercises. It is possible, though, that it will take several years to observe some of the critical thinking skills that were acquired as a result of the writing program.

The following is presented as a practical guide for mathematics teachers who are ready to use writing in their classrooms. Techniques will be presented from writing programs such as Writing across the Curriculum and Writing for Learning and from the domain-specific literature found in the mathematics educational journals. The techniques divide into two general categories: the shorter-term (less than a class period) techniques and the longer-term, more formal techniques.

Examine both categories of techniques, and, as the program develops, select and adapt those techniques that are appropriate to your grade level, teaching style, and students. At the start of the writing program, use one or two of the shorter-term techniques.

Shorter-Term Writing Techniques

These techniques may require just a few minutes of class time each day, or they may require the whole period. On any given day, either of these extremes may be appropriate. Begin by selecting one method that you feel comfortable with and adapt it to your classroom and your style.

Explain How

This procedure is similar to the pair problem-solving method except that each step is written rather than spoken. Students will need some practice to develop this technique, but it is well worth the effort. See the *General Writing Instructions* and *Writing Probes* sections at the end of this paper for ideas to help students become more proficient in this technique. Discuss the *General Writing Instructions* with students at the beginning of the writing program. Encourage them to use the *Writing Probes* questions whenever using this writing technique. J. Salem refers to this technique as "chatter"—"Students talking to themselves on paper as they do a problem." As an example, consider that solving an equation is undoing what has been done to a variable. Write what has been done to the variable in the following equation. Then state what must be done to reverse that.

$X/3 + 7 = -13$	"X is divided by 3, then 7 is added."
Subtract 7	This is the first step to undo that.
$X/3 = -20$	"X is divided by 3."
Multiply by 3	to undo that.
$X = -60$	

A similar technique is to ask students to write the steps to a solution. Non-mathematical problems work as well as mathematical ones here. For example, ask students to describe how to build a tower using blocks of different sizes and colors. Then have the students exchange their solutions and attempt to rebuild the tower using only the given instructions. Finally, have the students evaluate their solutions by discussing the results of the tower construction. The same procedure can be used with a mathematical solution.

Another alternative to this technique is to require that only the solution steps be written out; i.e. leave out all reference to the numbers and/or variables. For example, explain how to add $\frac{1}{3}$ and $\frac{1}{4}$ without mentioning $\frac{1}{3}$ or $\frac{1}{4}$. It is usually a good idea to add something to the instructions, such as "Assume you are explaining this process to a sixth grader." The "explain how" technique is a good one to start the writing program with.

Compare Two Concepts

In this technique the students are first required to define in their own words the meanings of various terms. They may start by writing rather poorly written paragraphs. As they are able to express themselves more precisely, the paragraphs will shorten to one or two sentences. After students are able to explain the meaning of individual terms in one or two sentences, give them pairs of terms representing either similar or contrasting concepts. For example, explain how the following concepts are related: equation-graph, line-plane, number-line coordinates, point-line, linear equation-quadratic equation, fraction-decimal number, circle-ellipse, function-relation, and geometry-algebra. When starting to use this technique again, the students may need to write a paragraph or more. Discourage the use of memorized definitions and concepts. As the technique develops, students should be able to formulate ideas more succinctly, and the paragraph will be reduced to one or two sentences.

Explain Why

This is a more advanced technique and requires more preparation on the teacher's part. The student should have done some form of writing, for example, the "explain how" method, before using this technique. This technique uses the open-ended question format, so that a final, definitive solution is never possible. After the students are able to explain how to add two fractions with some degree of proficiency, ask "Why would you do it this way?" Plan this activity carefully and be prepared for some chaos.

Word Problems

Although there are those who believe that "doing word problems is problem solving," there is much more to problem solving than this. Once the algorithm is acquired for a particular type of "problem," it is no longer a problem, but rather an exercise. Problem solving is no longer occurring at this point. There are many good references for word problems and for methods of generating problems appropriate to the goals of the instructional unit and to assure that problem solving is taking place. (For example, see A. Whimbey and J. Lockhead's *Problem Solving and Comprehension* and The National Council of Teachers of Mathematics' *Problem Solving in School Mathematics*.) The students should be required to have complete headings for each problem solution. Each item in the heading should contain the words *number of*. Rather than write *nickels*, they should write *number of nickels*, or *number of cents* represented by the nickels. Draw diagrams whenever possible. As the students acquire

201

the skills to "solve" a particular type of problem (i.e., they have learned the algorithm), give them other problems using the same solution method, but varied enough to require some additional thought process. For example, make the solution a part of a larger procedure, change the "setting," or add extraneous information.

Outline the Chapter

Use this technique for introducing new material or for reviewing for a test. This is a good way for students to learn to organize information. You might need to give some instruction in outlining methods.

Test Questions

For students to write test questions requires not only a good understanding of the material involved, but also the ability to organize and synthesize information. When the student has formulated a tentative problem, it must be evaluated to determine its suitability as a test question. This technique, utilized at the end of an instructional unit, brings together the knowledge of the unit at a high level of mental processing. Many of the higher-level cognitive processes are involved. Problem solving and cognitive learning are taking place. Begin this technique by dividing the chapter or chapters into parts and assigning each part to one or two students. Ask each student to prepare two or three problems. Finally, evaluate the questions, require rewriting as necessary, and compile the questions into a review test and the official test. This provides not only a meaningful learning experience for the students, but also problems for unit tests, final exams, and practice tests. The students need writing experience for this technique. However, it should provide an effective method of review and a valuable learning experience for a major unit in the mathematics program.

Replace Two-Column Proofs with Prose

The rigor of the formal two-column proof in geometry often causes students to develop a mental block that makes them unable to complete a proof. The prose form allows the students to write out their thoughts (and frustrations) and to examine and evaluate them. Because the thoughts are written down, they do not have to be held in short-term memory. This "unloading" of memory makes room for more pertinent information to be activated. After some experience with the prose form, students are usually able to organize information more effectively and meet with more success with the two-column proof.

Notebook

The notebook is really not a separate technique, but rather the use of several of the above techniques in a more organized way. The notebook has two primary purposes: for taking daily class notes during the class period and as a journal for keeping a personal record of thoughts and reflections at the end of the period. At times the two overlap. For example, students may be asked to write out all of the steps of a problem solution in the notebook during the class period. However, at the end of the period, they may express in the journal their displeasure (or pleasure) at having to write out all of the steps of the problem solution.

• *Daily Notes.* This is where much of the writing for the above techniques takes place. The teacher should be posing problems at the proper level for learning to be effective. The solutions should be recorded in the notebook as examples. The lecture notes and notes on any class discussion should be rephrased in the student's own words and recorded. One successful method is the teacher-led class discussion to arrive at a definition of an important term. The class discussion continues until the appropriate level of precision in the definition is obtained and is understood by the class. Vocabulary lists with synonyms and definitions derived in this fashion should be recorded.

• *Journal.* The last five to ten minutes of each class period should be devoted to writing in the journal. This time for writing is usually for reflection on what was learned in class, on a difficult concept that needs some attention, on applications of the day's material, and on feelings about the material, the teacher, or the classmates. You might ask students to write about a critical concept. For example, after completing a unit on factoring, ask the students to discuss the following statement: "Factoring and finding a product are reverse processes." Journal writing gives students time to think at their own rate and to internalize new concepts by relating them to their own experiences. As students become familiar and comfortable with journal writing, their responses will show definite improvements in organizational skills. As a result, their work becomes easier to read. The students should also be able to ask the teacher questions in the journal that may be difficult to ask in class.

Longer-Term Writing Ideas

After the writing program is well established, the teacher may want to assign a writing project that will require several days or weeks to complete. The project should be carefully designed to assure that it meets the course/unit objectives and is within the students' technical and writing capabilities. Several of these projects require access to libraries, banking information, and other records. The following is a list of ideas for longer-term projects.

- Paper on a career in a field related to mathematics.
- Book report on a mathematics or computer book or paper; e.g., *Mindstorms*, papers from *The Mathematics Experience*, *Men of Mathematics*, or *The World of Mathematics*.
- Write a letter to a national testing service, offering suggestions on methods to evaluate and measure intelligence.
- Develop a document outlining a plan to form a financial planning group that would give advice to students on how to earn, save, and manage money.
- Report on statistics as used by insurance companies.
- Paper on a famous mathematician.
- Report on mathematics anxiety.
- Report on how mathematics is used to measure and evaluate intelligence, reading skills, mathematical abilities, and other aptitudes.
- Report on the historical development of some area of mathematics, such as algebra.
- Choose a chapter from a mathematics book on a topic not covered in the course this year. Study it, keeping a journal of your thoughts and feelings as you progress through the unit. Summarize your journal in a report.
- Report on new development in computer technology.
- Letter to a friend on how he or she feels about mathematics.
- Letter to a friend offering advice on solving a certain problem.
- Discussion of a retired person's budget, considering various expenses and sources of income.
- Present a longer mathematics problem solution. Write out the solution in words (no mathematics), describing each step with its reason.
- Report on the origins of a particular measurement; e.g., length of a day.
- Take the role of a lawyer, dentist, stockbroker, etc., and write a letter explaining why one should master the concepts in Algebra I.

WRITING PROBES

Ideas to help you form your thoughts as you are writing.

General Probes
- What are you thinking?
- Can you say more about that?
- Explain that so that a fifth grader would understand it.
- Write that again in a different way.
- Are you sure about that?
- Check the accuracy of that last step.
- What were you thinking about to get that?
- What did you mean by "it"?
- Label that step so you can refer to it.
- What does that X stand for?
- What does that equation say in plain English?
- Could you draw a picture of that?
- Why did you write that equation?
- Tell why that last step makes sense.

When You Are Stuck
- Why are you stuck?
- What other information do you need to get unstuck?
- How can you get needed information?
- Do you need to go back and review some of your steps?
- What makes this problem difficult?

When You Think You Are Finished
- Do you have other ideas that you have not written about?
- What were you thinking about when you did step X?
- How can you show that you did this correctly?
- Could this have been done another way (more eloquently)?

GENERAL WRITING INSTRUCTIONS

- Try to write down everything you are thinking.
- Write down each step as if you were explaining your thoughts to a friend.
- If you have done pair solving, imagine the questioner sitting next to you, asking you to explain each step.

- Your writing is not right or wrong; it is what you are thinking at that time.
- Don't worry too much about spelling, but try to make your writing grammatically correct so that others can understand it.
- Rephrase thoughts so that you will be sure others will understand them.
- Draw diagrams whenever possible.
- Describe preliminary ideas, discuss why you accepted or rejected each one, and then explain how you expanded on those you kept.
- When you have completed your work, go back and review it. Record your final impressions.

OTHER T&W PUBLICATIONS OF INTEREST

The Story in History: Writing Your Way into the American Experience by Margot Fortunato Galt. Practical exercises give students and teachers of all levels an entirely new way to view history—by reexperiencing it from the vantage point of the imaginative writer.

Like It Was: A Complete Guide to Writing Oral History by Cynthia Stokes Brown. This how-to guide was written by a teacher who won the American Book Award for her work in oral history. For students 12 and up and for English, social studies, and history teachers. "A solid, well-or⸱nized introduction, covering everything" —*Booklist.*

Personal Fiction Writing: A Guide for Writing from Real Life for Teachers, Students, & Writers by Meredith Sue Willis. "A terrific resource for the classroom teacher as well as the novice writer" — *Harvard Educational Review.*

Origins by Sandra R. Robinson with Lindsay McAuliffe. A new way to get students excited about language: the fascination of word origins. Includes many writing exercises and a brief history of English. "Refreshing and attractive" —Robert MacNeil.

The Teachers & Writers Handbook of Poetic Forms, edited by Ron Padgett. A clear, concise guide to 74 different poetic forms, their histories, examples, and how to use them. "A treasure" —*Kliatt.*

The Writing Workshop, Vols. 1 & 2 by Alan Ziegler. A perfect combination of theory, practice, and specific assignments. "Invaluable to the writing teacher" —*Contemporary Education.*

The Whole Word Catalogue, Vols. 1 & 2. T&W's best-selling guides to teaching imaginative writing. "*WWC 1* is probably the best practical guide for teachers who really want to stimulate their students to write" —*Learning.* "*WWC 2* is excellent . . . Makes available approaches to the teaching of writing not found in other programs" —*Language Arts.*

•

For a complete catalogue of T&W books, magazines, audiotapes,
videotapes, and computer writing games, contact:
Teachers & Writers Collaborative
5 Union Square West
New York, NY 10003-3306
(212) 691-6590